Animals of Ohio's Ponds

and Vernal Pools

Animals of Ohio's Ponds and Vernal Pools

David FitzSimmons

Photographs by Gary Meszaros

The Kent State University Press

Kent, Ohio

© 2011 by The Kent State University Press, Kent, Ohio 44242

All rights reserved

Library of Congress Catalog Card Number 2010035821

ISBN 978-1-60635-081-2

Manufactured in China

Library of Congress Cataloging-in-Publication Data

FitzSimmons, David.

 Animals of Ohio's ponds and vernal pools / by David FitzSimmons ;
photographs by Gary Meszaros.

    p. cm.

 Includes bibliographical references and index.

 ISBN 978-1-60635-081-2 (hardcover : alk. paper) ∞

 1. Wetland animals—Ohio. 2. Ponds—Ohio. 3. Vernal pools—Ohio.
4. Wetland ecology—Ohio. I. Meszaros, Gary, ill. II. Title.

 QL198.F58 2011

 591.76809771—dc22

               2010035821

British Library Cataloging-in-Publication data are available.

15 14 13 12 11 10   5 4 3 2 1

For Olivia, Sarah, and Phoebe, whose encouragement

is as profound as their love of nature

DF

To Jane and my daughters, Amy and Carrie

GM

Contents

Acknowledgments

I would like to thank the following people, without whom this book would not have been possible. First, I am grateful for my loving, patient, and thoughtful wife, Olivia, whose assistance and encouragement provided the supportive environment necessary to complete such a project. I am thankful for my daughters: Sarah, whose gleeful trips to wonderfully watery worlds—splashing in the mud, catching critters, always asking questions—symbolize why this book may benefit future generations; and Phoebe, whose bright infant eyes, ebullient smile, and natural curiosity clearly reflect the vivacity, splendor, and goodness of our world. I am grateful to my parents: my mother, Judy, for her skills in language arts and love of the natural world, and my father, Mick, for his experiential approach to natural history and enjoyment of the world. Both of them catalyzed my commitment to appreciating nature and sharing its spirituality. And I would not be who I am without my brothers: Jim, whose mathematical precision, grammatical acumen, and appreciation for the natural world keeps me toeing scientific, rhetorical, and aesthetic lines; and Tom, whose artistic and playful skill with words, sense of humor, and humility hopefully inspire vital writing imbued with healthy perspective. In a way, this book is a family travelogue.

I am extremely grateful to Gary Meszaros, who asked me to join him in his decades-long project, especially for bringing me close to his visions and knowledge of the natural world. Gary's passion for natural history is only matched by his energy for producing another and yet another beautiful, moving image. Thank you, Gary, for expanding my photographic repertoire.

I appreciate Kent State University Press, especially director Will Underwood, for thoughtful guidance and careful attention to the details of quality publishing. I am grateful for the assistance of Tim Matson and Ralph Pfingsten, who expertly reviewed the manuscript. Thanks to Ashland University for support during the research and writing of this book, especially to my colleagues and friends who provided ideas and encouragement, namely Dan Lehman, Deborah Fleming, Joe Mackall, Soren Brauner, Richard Stoffer, Karen Stine, and Doug Kane. I am grateful for the support of the Sigma Corporation of North America and the ever-encouraging Dave Metz. Thanks to my enthusiastically supportive friend, Matt Mitchell; naturalists Steve McKee, Merrill Tawse, Guy Denny, Bill Fisher, Warren Uxley, and Robert Glotzhober; photographers and friends Ian Adams, Randall Schieber, and Art Weber; the Ohio Department of Natural Resources, especially Nancy Strayer, Louis Andres, Jason Wesley, Jim McCormac, Mort Pugh, and Mike Wilkerson; Lisa Barnese-Walz of the U.S. Army Corps of Engineers; Adam Hater of Jones Fish Hatchery; and Julie Zickefoose.

Finally, I thank God for a fascinating natural world, full of complexly beautiful environments and incredibly amazing creatures. And I remain ever grateful for opportunities to share my experiences with others.

Gary Meszaros wishes to thank Ian Adams, Jim Bissel, Guy Denny, Bob Glotzhober, Tim Matson, Jim McCormac, Jane Meszaros, Dwight Moody, Ralph Pfingsten, Carolyn Platt, John Pogacnik, Dan Rice, Ed Stroh, Jeff Wolfinger, and Jeff Riebe.

Introduction

Why Should We Care?

Water is everywhere in Ohio. Myriads of streams course the state, and lakes, ponds, and pools ubiquitously pepper the topography. As humans, we have an interest in water. It not only provides us with recreation but also is necessary for our survival. Indeed, water comprises over 90 percent of our body mass.

But we also need water in a deeper, more spiritual way. Water sustains our bodies and our souls. Somehow hardwired into our very beings are connections to our aquatic roots, our earliest ancestors who lived and later emerged from the earth's waters. In a way, this book traces that lineage. The chapters here follow groups of animals in their evolutionary order, helping redraw the lines of descent from primitive, primordial water dwellers to more complex and recent land dwellers.

Ponder this: How many days go by when you do not see some form of water—a river, creek, pond, or pool? The answer is not one day. In fact, if you do not get your fill of water—not just drinking it but also experiencing it—then your inner self is probably beckoning for more. First of all, there is something relaxing about water, whether hearing the lapping of waves, canoeing down riffling rivers, or experiencing raindrops pitter-pattering across a pond or pool. All of these sounds are forms of music that replenish, invigorate, and fulfill the soul. And mixed with this symphony are other instruments, other organisms completing the orchestra. These fellow creatures are all one way or another our relatives, animals who evolved with us from earth's earliest life forms. Some never left the water. Some abandoned it for good. And still others departed from waters and eventually returned.

Animals of Ohio's Ponds and Vernal Pools is dedicated to this returning, helping us find ways to discover again parts of ourselves that are worthy of more exploration. The ponds and vernal pools discussed here are not just "water," and the animals examined in this book are not just "things." Ponds and vernal pools are specialized niches in our environment, their ecology worthy of our study. And the inhabitants of these ecosystems—frogs, fish, snails, salamanders, and countless other animals—are all different, each deserving our closer inspection. Just labeling "creepy" things as "bugs" does a disservice to them. We are sophisticated enough to discern the various species, distinguishing, for example, a true bug from a spider. Humans are part of the biological nexus, and we improve our chances of survival and increase our enjoyment of life by knowing a giant water bug from a six-spotted fishing spider. Which of these packs a walloping bite? Which is more dangerous? The answers to these questions are not at the back of this book: they are within . . . within you to discover, in the pages of this book and in nearby ponds and vernal pools.

We as humans tend to think highly of ourselves. We arrogate and dominate. But can we walk on water? Can we breathe through our skin? Can we catch our dinner at 30 miles per hour? Can we spend an entire winter under water in frozen animation? Do we know how to travel by foot, in the rain, on a rainy night, to an exact location, year after year, without a GPS?

While *Homo sapiens* can design incredible skyscrapers, paint beautifully impressionistic water lilies, find cures for diseases, and pluck the strings of instruments in mesmerizing ways, there is much that we cannot do. Humans are amazing; nature is equally so. This book, first of all, provides a way to continue your journey of better understanding the wildly complicated and amazing world around us by focusing on the fascinating fauna of ponds and vernal pools. Indeed, this book could just as easily be about fungi in Ohio's woodlands or grasses in the state's prairies. But it is ponds and vernal pools now; in the end, however, this book will only be an introduction to a variety of ecosystems, all well worth exploring and embracing.

This leads to this book's second purpose: incorporation. After exploring and learning about the world around us, the process has only begun. True humanistic learning involves not only gathering information but also adjusting our worldviews to include new discoveries. Witnessing unbelievable sights, sounds, smells, tastes, and feelings may overwhelm our senses at times, but this is how we embrace the experience that is the complex web of nature. Let the experiences of sitting pondside on a sunny summer day or wading in a vernal pool on a cool spring night fill you and sustain you. This, of course, takes practice. But, once we see an aquatic caddisfly larva build its square home, push water through it for more efficient breathing, and later metamorphose into an entirely different-looking flying organism, we cannot help but do the same: we change into individuals who see, think, and behave differently. We, too, begin to soar.

This leads to the final purpose of this book: environmental conservation. Climate change is firmly upon us. Aquifers are drying and habitats are disappearing. To be sure, change is the way of life, but not all change is good, as global warming from carbon emissions points out; we can, however, reason and respond. Rampant urban sprawl, poor forestry practices, uncontrolled or undercontrolled pollution, grossly enormous carbon emissions, and other results of poor human decisions—all attributable to the most advanced mammal found in and around our wetlands—are ruining these same waters and erasing their inhabitants. As we destroy these worlds, we destroy part of ourselves; the most pressing purpose of this book, therefore, is not environmental awareness but rather getting people to cherish the world around us, encouraging heartfelt stewardship of our natural treasures.

If we do not revise our individual behaviors, rethink our community decisions, reinvent our businesses, reformulate public policies, and repair ourselves spiritually by remembering our evolutionary origins, the animals of Ohio's ponds and vernal pools will suffer. Of course, not only will the animals suffer; so will you.

Not Just a Fishin' Hole

Farm Ponds and Other Diminutive Lakes

N early anywhere you go in Ohio, you are bound to find ponds. Whether it is a natural pond left by a glacier or man-made farm ponds across the state, you do not have to look hard to find a variety of animals: fish swimming, frogs croaking, herons hunting, muskrats diving, insects flying, bats drinking, and, of course, humans enjoying. In the Buckeye State, water is generally abundant, and the varied topography of the state lends itself well to many kinds of ponds and an abundance of animals within and around them.

What Is a Pond?

Around the world, bodies of water vary from the largest oceans to the smallest puddles. In Ohio, we do not have large bodies of salt water; we do have, however, all kinds of freshwater bodies, ranging from Lake Erie to smaller reservoirs, ponds, and pools. Further, the state is coursed by waters such as the majestic Ohio River as well as a variety of smaller rivers and streams. From this wide array of Ohio waters, how do you distinguish a pond from other wet habitats?

First of all, it is useful to make a distinction between *lotic*, or moving waters, and *lentic*, or still waters. Lotic waters include anything from raging rivers to crayfish-filled creeks. The brook behind your house or the large river into which it flows are all lotic waters. These moving waterways are not the subject of this book, although many of the animals described in subsequent chapters may be found in rivers and streams or along their banks.

Lentic waters can range in size from as large as Lake Erie, Buckeye Lake, or Grand Lake Saint Mary's to as small as placid ponds, vernal

Opposite: A spreadwing damselfly nymph is on today's menu for this red-spotted newt. Newts in turn provide meals for snakes and turtles.

pools, and even minute mud puddles. Ponds are the middle group here, having less surface area and less depth than large lakes but more surface area and greater depth than pools and puddles. In short, we may say that ponds are *small, permanent, shallow bodies of standing freshwater with relatively small temperature gradients from top to bottom.*

While to some scientists a pond is just a "small lake"—that is, the term "pond" is more of a popular description than a scientific distinction—some subtle differences do exist between larger lakes and smaller ponds. To begin with, ponds are so small that wave formation is not possible. On larger lakes, wind blowing across the surface can begin the physical process of creating peaks and troughs in the water's upper layer. Waves form, and they travel across the lake, causing the lake's edges to have exposed rock, thick mud, or strands of beach, all unique environments for plants and animals. Ponds are small enough in size that, while winds may blow across them, their surface areas are not large enough for waves to form. During a strong wind you might see ripples forming on a pond's surface, but you will never see sustained waves that roll into the shoreline and erode the land.

Ponds are also permanent. This means that each one maintains a relatively constant water level. From season to season—and especially from wet years to dry years—there may be some filling or drawing down but not a lot. The watershed of a well-designed farm pond, for example, is sufficiently large enough to keep its basin filled year-round, and extra precipitation often will exit the pond from a carefully designed outlet. Many ponds have substrates, such as clay, which assure that water is held throughout the year.

Further, ponds are shallow bodies of water. They lack the depth found

Floating duckweed, one-eighth inch across, shares the water surface with pondweeds, spatterdocks, and water lilies.

in larger lakes. A good measure of pond depth is based on light transmission and plant life. Generally, ponds are shallow enough to allow light penetration across their entire basins, all the way to the bottom. This means that ponds may have plants living from shore to shore, even in their deepest waters. Of course, this can vary within a pond due to the *turbidity* of the water. Turbidity is the measure of particles and other suspended matter in the water, which causes reduced transmission of light. Sometimes water may become so clouded with organic or inorganic materials, such as *detritus* or silt, that light transmission is limited greatly and plant growth becomes difficult. Ponds, then, are shallow enough to allow plants to grow in their deepest parts, yet inflows of particles may increase a pond's turbidity and limit plant growth periodically.

The final determining factor for whether small bodies of water are considered ponds has to do with temperature. Generally speaking, ponds do not vary greatly in temperature from top to bottom; larger lakes, however, tend to have significant differences in water temperature from the top to bottom. For example, on a summer day, the surface of Lake Erie may be almost thirty degrees (Fahrenheit) warmer than the water at the deepest parts. And readings taken by the Army Corps of Engineers at Caesar Creek Lake, one of the deepest reservoirs in the state, show similar summertime differences. Ponds, on the other hand, tend to be quite shallow and, therefore, do not have such extremes.

As a general rule, then, it is safe to say that ponds tend to be around twenty feet or less in depth and not much bigger than a few acres in size. If the water is much deeper, the plants cannot survive on the bottom, and

Left: Giant water bugs feed largely on tadpoles and other small animals. They have powerful front legs for grasping and holding victims.

Right: Water striders have the ability to propel themselves by altering the surface tension of the water. They capture prey by watching for changing light on the surface.

if it gets much larger in surface area, then winds can create waves, creating different habitats along the shoreline than found along ponds.

You can probably see now that precisely defining a pond is difficult. Lake Erie is obviously a gigantic lake, and the half-acre body of water near the barn down the road is surely a pond. But what do you call waters that are four acres in surface area and twenty-five feet deep? Are those just ripples forming on its surface, or are there sustained waves? Are plants able to grow in the deepest sections? If you go swimming, do you find the water at the bottom significantly cooler than the water at the top? These questions help point out that the progression from a small pond to a large lake is along a continuum.

Pond Formation

A tour of Ohio's thousands of ponds reveals a variety of ways they have been formed. The origins of ponds in Ohio range from the catastrophic to the domestic. During the Pleistocene epoch, which stretched from about 1.8 million years ago to about 10,000 years ago, four glaciers acted as early bulldozers, scouring our landscape and creating Lake Erie and its Great Lake cousins. But it was not so much the scraping and pushing action of the earlier three ice sheets that formed most of Ohio's naturally formed ponds. Many of our glacially formed ponds are the result of the fourth and most recent glacier in Ohio, the Wisconsinan, which left the state about 12,000 years ago.

During one of the last cold periods of the Pleistocene epoch, the Wisconsinan glacier advanced into Ohio, fed from ice formed in central Canada. It advanced through much of Ohio, changing the topography. Eventually the climate changed, and the earth's atmosphere warmed. Consequently, the Wisconsinan glacier began to melt, retreating from Ohio. Some of our ponds are remnants of filled-in lakes that were originally dug out by the advancing ice; others formed when chunks of ice fell off the front edge of the receding glacier. Ice broke off the southern edge of the glacier and sat upon or lay buried within the silt, sand, and rock debris left from the glacier. This till and outwash surrounded the block of ice, which slowly melted, often over a period of a hundred years or more. In the end, the space left by the ice became depressions that filled with water. These became *kettle holes*. Many kettle holes are lakes, bogs, or just swales in Ohio's topography; some of them constitute ponds today. An airplane ride over the parts of Ohio that were previously glaciated, especially the flatter sections of the northeast, reveals

This female pondhawk has captured a forktail damselfly. Dragonflies have powerful jaws for tearing apart prey.

many gradually sloping ponds, often circular or ovoid, long ago formed by glacial ice blocks. Two easily accessible glacially formed ponds are Stage's Pond in Pickaway County and Lake Kelso in Geauga County.

A smaller percentage of ponds in Ohio formed through other not so catastrophic geological forces. Some ponds in the state came about not from frozen water but from running water—that is, from rivers and streams. Commonly, waterways change their courses, creating unused channels that may become basins for ponds. Abandoned oxbows of rivers are excellent examples. Over time, rivers will cut through thin earthen walls separating the narrowest parts of oxbows, taking the paths of least resistance. The straighter-flowing rivers then cut lower and lower into their valleys, leaving the oxbow depressions to fill with precipitation alone. These lentic environments can become permanent ponds. A number of oxbow ponds have formed, for example, along the Great Miami River above the confluence with the Ohio River.

Along Lake Erie, ponds can form in sand dunes. Waves carry sand to the shores, and, as the sand dries, winds pick up particles and move them inland from the shore. Slowly, dunes form inland from the water. Unstable dunes shift their positions over time as winds relocate them particle by particle. Sometimes winds blow just right, and deep depressions

GLACIAL MAP OF OHIO

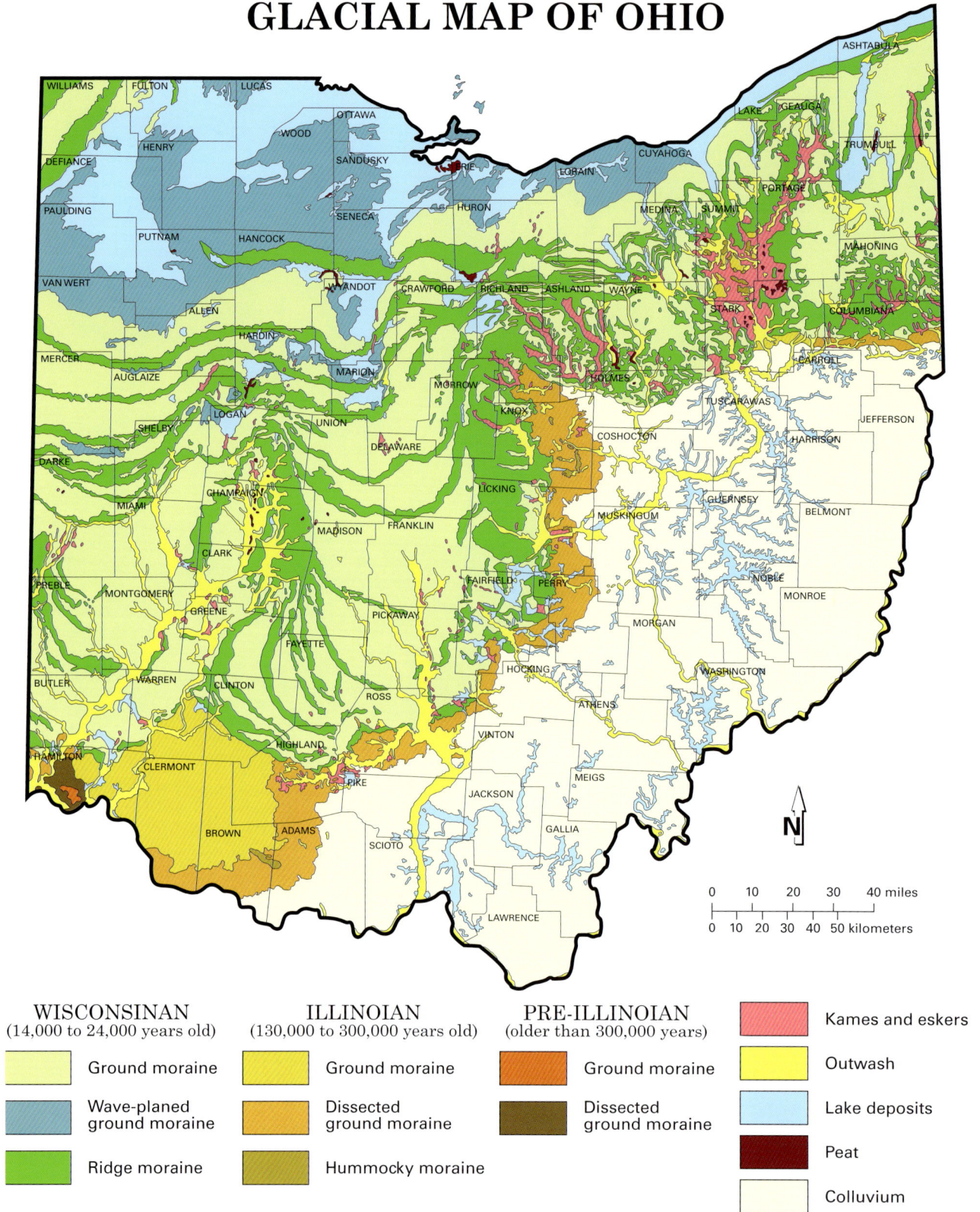

WISCONSINAN
(14,000 to 24,000 years old)

- Ground moraine
- Wave-planed ground moraine
- Ridge moraine

ILLINOIAN
(130,000 to 300,000 years old)

- Ground moraine
- Dissected ground moraine
- Hummocky moraine

PRE-ILLINOIAN
(older than 300,000 years)

- Ground moraine
- Dissected ground moraine

- Kames and eskers
- Outwash
- Lake deposits
- Peat
- Colluvium

Figure 1.1. Glacial map of Ohio. Courtesy
of Ohio Department of Natural Resources,
Division of Geological Survey.

form between one dune and the next. Precipitation collects in these basins, and ponds form. These dune ponds can form unique environments, hosting species of plants and animals quite different from those found in other Ohio ponds. Good examples of dune ponds exist in the Oak Openings region of northwest Ohio.

Many ponds are not formed through geological, hydrological, meteorological, or other forces of the environment but rather by animals. Until recently, the animal in Ohio most responsible for forming ponds was the beaver. The largest members of the rodent family and one of the few animals able to control their own environments, beavers cut down shrubs and trees, mix pieces of them with other plant material and mud, and dam creeks and streams, forming ponds. Beavers live in colonies—working, eating, and raising young together—in and around their specially designed ponds. The standing water behind beaver dams provides protection for their lodges and a permanently wet habitat for feeding and rearing young. While beavers were common before white settlement, extensive trapping resulted in the mammal's *extirpation* from Ohio in the early 1800s. Since then, the beaver has slowly returned to Ohio, becoming increasingly abundant. Population estimates from the Ohio Department of Natural Resources, Division of Wildlife show the number of beavers in Ohio approximately doubling in the last quarter century.

Over the past several hundred years, however, one animal in Ohio has far outpaced the activity of beavers in pond construction. That animal is, of course, *Homo sapiens*. We, as thinking, desiring, designing humans, construct ponds for a plenitude of reasons. We build ponds for watering farm animals and protecting our homes from fire. We construct ponds for recreation, such as swimming and fishing. We dig ponds for the aesthetic enjoyment of listening to spring peepers and bullfrogs or watching a sunset reflected in its waters. We even build ponds by accident, the outcome of earthmoving (borrow pit ponds) or mining.

The Hydrology of a Pond

Once the basin of a pond is formed, one thing is needed: water. But where does it come from? The answer is precipitation. Precipitation is the original source of all water in ponds, either directly or indirectly. While you might think that rain and snow falling directly into a pond itself is the major source of the water, in fact, only a small fraction of a pond's volume comes from direct input. Instead, most of the water that reaches a pond comes from surface runoff and groundwater seepage. When atmospheric water falls as precipitation, it seeks the lowest point, and this gravitational attraction brings water from surrounding higher areas downhill into the basin of a pond. When the water flows down the hills of the watershed and into the pond itself, this is called surface inflow. When the water falls as precipitation and soaks into the ground, it may enter the pond as groundwater, often seeping from a hillside spring.

Outflowing waters can leave a pond in a variety of ways. The most obvious is through a drain or spillway. Farm ponds are often built with pipes as outlets. These determine the upper limit of the water level and provide consistent drainage. Beaver ponds are built with level tops or spillways over which excess water flows. Seepage through the basin is another way that ponds lose water. This type of outflow varies greatly according to the substrate of the basin. The more clay-rich a soil is, the better a pond will seal and the less water loss there will be due to groundwater seepage.

The last factors in pond outflow are *evaporation* and *sublimation*. Evaporation is the changing of water from a liquid state to a gaseous state (via the input of energy). During the hot summer months, the baking rays of the sun quickly provide the energy necessary to release large amounts of water from ponds in the form of water vapor. Even

Dytiscus diving beetles are voracious predators. They can inflict a painful bite if not handled properly.

during winter, enough solar energy is available for vapor to be lost from ponds. In this case, the water loss is not from evaporation but through sublimation, the changing of water from a solid state directly to a gaseous one (again with the input of energy). In both cases, more surface area translates into a greater potential for gaseous outflow. It follows, then, that shallow ponds have a larger percentage of their water loss attributed to vapor than deeper ones.

Hydrologically, ponds with inflow and outflow are referred to as *open*, while those with no outflow (besides evaporation and sublimation) are *closed*. Habitats for animals can vary widely depending on this open or closed hydrology. Inputs of chemicals from the surrounding watershed can create poor habitats for some organisms and excellent ones for others. In closed systems, these chemicals tend to collect and become concentrated. Nitrogen fertilizer, for example, can enter a pond through surface runoff and create algae blooms, boons for alga-eating animals such as snails, but the algae also crowd out light and deplete the level of dissolved oxygen in the water, effectively killing off other organisms, such as fish. Basins with little water input and no drainage can create stagnant situations where acids collect and oxygen levels are greatly depleted, an environment in which many plants and animals cannot survive.

Flying above the pond surface, twelve-spot skimmers are a common sight.

The Energy Cycle in Ponds

Ponds obviously depend upon spring rains, winter snow, and other precipitation all year long, but less obvious is their reliance upon the sun. When it comes to powering pond organisms, the sun is central. In fact, solar radiation is the fundamental source of energy for ponds, as for all aquatic ecosystems. The specific pathways by which solar energy reaches pond organisms vary. On a stroll beside an Ohio pond on a sunny summer afternoon, you would likely see hundreds of plants along the shore and throughout the pond—cattails, sedges, pond lilies, duckweed, watermeal, and large colonies of algae. This might lead you to believe that these diverse and multitudinous plants are the largest factor in capturing the sun's rays and converting them into usable energy in the pond.

In fact, plants are major players in converting solar energy into organic compounds usable within the entire pond ecosystem, but recent studies show that most inland aquatic ecosystems depend more upon external inputs of energy than upon production from within. All ecosystems—including ponds—lie along a continuum from creating their own energy (*autochthonous* production), namely from plants photo-

Ohio has thousands of ponds, mostly man-made. Many recently constructed ponds have been created with wetland mitigation funds.

The water boatman's hind legs serve as oars to propel it through water. Water boatmen consume bits of plant debris.

synthesizing, to deriving their energy from outside sources (*allochthonous*), such as from the windfall of leaves or from animal waste dropping into the pond. While it is understandable that ponds are not entirely autochthonous, it is not so obvious just how much ponds are deeply allochthonous; indeed, studies show that in inland lakes and streams, the vast majority of the energy input comes from detritus, usually in the form of dissolved organic matter. While ponds are not entirely allochthonous, they depend greatly on the photosynthesis of plants both inside and outside their basins. Moreover, some ponds are surrounded by trees and therefore have only small amounts of photosynthesis occurring. These woodland ponds are almost entirely dependent upon falling leaves and other outside sources for their energy input.

Understanding the movement of power from the sun to usable forms of energy for organisms in pond waters requires looking closely at the three major types of organisms in ponds. First, there are the *producers*. These organisms are mainly responsible for solar energy conversion. They perform *photosynthesis*, the cellular-level process by which sun energy is captured for organic use. As sunlight falls on the *chloroplasts* containing cells of producers (most of which are plants), carbon dioxide and water are converted first to glucose and then to other energy-rich sugars, starches, and other carbohydrates. The sun's energy is thus harnessed, converting simpler compounds into energy-storing organic ones. This energy is stored by the producer until released in the biological processes of living—namely during the creation of new organic compounds, in transporting compounds within the organism, and in the case of the rare animals that can carry on photosynthesis, in locomotion and even bioluminescence.

In many pond ecosystems, however, the majority of this primary production is not carried out in the water itself; rather, the majority of the photosynthesis occurs beyond the pond, and the energy that is produced is carried into the pond in the form of detritus. Dead animals and plants, broken-off tissues, and waste materials are the constituents of detritus. Detritus is nonliving organic matter that is broken down by the second category of organisms: *decomposers*. Nature's waste-clean-up organisms derive their energy by breaking down organic leftovers, and in the process they also return important chemical constituents to the ecosystem as they decompose more complex organic matter, leaving behind them inorganic compounds that future producers may use.

Purple pickerelweed spires adorn Ohio wetlands throughout the summer.

Opposite: Submerged and covered with duckweed, a bullfrog surveys its surroundings.

Diving beetle larvae, like their adult counterparts, are armed with enormous jaws to capture and hold prey.

It should be noted that decomposers are not the only avenue for organic substances to be returned to simpler, inorganic states. Within ponds, organic matter can be broken down by ultraviolet light. This process, know as *photolysis*, is another way that the molecules of organic matter can be broken down and rearranged into inorganic compounds in the continual cycling of nutrients within ponds.

The last link in the movement of energy within ponds is the group of animals that feeds on other organisms. These *consumers* are wholly dependent upon other organisms to capture solar energy, for these grazers and predators acquire energy-rich organic matter through the ingestion of other living tissues. In short, consumers eat producers and decomposers. A snail eating a plant leaf is an example of such an energy transfer. Of course, this energy can be transferred even further as one consumer eats another (e.g., a water bug eating a snail, a fish feeding on the insect, and a red-tailed hawk devouring the fish).

The first level of consumers, which eat plants, are called *herbivores*. They are considered *primary consumers*. The organisms that eat these primary consumers are called carnivores, and they are considered *sec-*

ondary consumers. Organisms that eat secondary consumers are tertiary consumers, and so on. The continuation of this food chain can go on for several levels; as this process continues, however, little of the energy is transferred from one level to the next. Inefficiently along each step, much energy is lost—directly to heat as well as in maintaining the day-to-day functions of the organisms. In fact, as one organism eats another from a lower level, only about 10 percent of the energy is transferred upward.

The entire process of nutrient cycling is entirely dependent upon sunlight reaching the photosynthetic organisms in and around the pond. Yet many factors mitigate solar radiation reaching the pond and its environs. Weather, seasonal day length, atmospheric pollution, and altitude can determine how much light may reach the earth. For production within the pond ecosystem, generally only about 5 percent of the sun's light energy reaches the producers.[1] The amount of light that penetrates the water of the pond itself is much less. The sun's ability to reach down into the depths of a pond is limited by a variety of factors, including dissolved or suspended inorganic and organic compounds in the water (*turbidity*), the angle of sunlight hitting the surface of the water according to season and latitude, ice and snow cover, and, of course, organisms (namely plants) growing in, on, or above the water.

Note

1. Wetzel, *Limnology*, 53.

Springtime Gems

Vernal Pools

As vital as ponds are, nothing exceeds the activity of a vernal pool on an early, rainy spring evening. Around the Ides of March, hundreds of thousands of chorus frogs, wood frogs, and spring peepers, along with their amphibian relatives, mole salamanders, flock to small depressions that fill with water from melting snow and falling rain. There they gather to dance, sing, and mate.

Smallish male and egg-laden female wood frogs congregate in shallow, cold pools—hundreds upon hundreds, sometimes thousands—vociferously clacking, attracting a partner, and then joining together in the connubial clasp called *amplexus*. Dispersed all around the newly joined couples are the fruits of the wood frogs' labors—transparent egg masses, jiggly looking collections of lentil-sized eggs, the beginning of future *Lithobates sylvatici*, most of whom two to three wet springs later will return to the same pool to repeat this fantastic calling, partnering, and teleological union.

As if that were not enough activity in a space sometimes as small as a one-car garage, thousands of mole salamanders (*Ambystoma* spp.)—the most common of which are spotted salamanders, Jefferson salamanders, and tiger salamanders—migrate to the same pools where the wood frogs collect. Across leaves and sticks, their slimy skin kept wet by the first warm rains, salamanders hone in on their ancestral breeding pools. The silent salamanders begin courting dances, swimming in sidewinding paths up from the bottom of the pool, breaking the surface with their snouts, and descending back down to the bottom. Over and over the males perform this slithering signal, a message to the congregating females. On the bottom of the pool, the male *Ambystoma* leave tiny white *spermatophores*. These white deposits are about the size of peppercorns and contain the advertising male's sperm, which the

Opposite: Female American toads lay long strands of eggs in the spring.

Vernal pools often dry up by late summer.

attracted female will pick up in her *cloaca*. Then she lays masses of eggs, attaching gelatinous globs onto plant stems, branches, and leaves that lie in the pool.

After remaining in the pool for a few days to a few weeks, the amphibians make their tortuous way back to their burrows. Frogs leap back into the surrounding woods and hide below leaf litter, under logs, or in holes. Similarly, the aptly named mole salamanders bury themselves, seldom seen throughout the summer, fall, or winter. But come next spring, both will be highly visible, returning to spring-filling pools to continue the life cycle—finding partners and creating new life for the pool and the surrounding woods and fields—although for any given year, only part of a local population migrates, which helps ensure species' survival.

But just as quickly as vernal pools fill with water and then amphibians, so too will the pool begin to dry up as summer progresses. The frog and salamander eggs will hatch, and countless tadpoles and salamander larvae will wriggle around the basin, seeking food—or becoming food for visiting reptiles, birds, and other predators, including their own kind. They will hurry to mature and leave before the water disappears completely.

As the spring progresses into summer, water begins to evaporate. By mid- to late summer—as tadpoles become frogs and *Ambystoma* larvae *metamorphose* into juvenile salamanders—the pools grow dry. Eventually, no water is left. Clearly, such desiccated locations are not suitable homes for fish and other predators, and this is exactly why certain species, such as wood frogs and mole salamanders, find vernal pools a perfectly matched site for reproduction and larval growth.

Symbols of wetlands, spotted salamanders return to their breeding pools in spring to mate and lay eggs.

While looking like small, shallow ponds, vernal pools are quite different. Most notably, vernal pools are usually temporary, while ponds are permanent. As their name implies, vernal pools typically fully fill in the spring (*vernus* in Latin refers to spring), but after topping out in late spring, they dry up, sometimes completely, in late summer.

Just mentioning the temporary nature of vernal pools, specifically their seasonal filling and drawdown, is not enough to define these unique ecosystems. If you think about it, collections of water in a farmer's field, wa-

A male American toad clasps a female in amplexus, waiting to fertilize her eggs.

ter pooling alongside the road, or even a large mud puddle at the neighborhood playground might fill with melting snow and falling rain and eventually dry up as the weather gets warmer. In general, we can identify all of these ephemeral bodies of water as *temporary pools*. In contrast to the definition of ponds from the last chapter—small, permanent, shallow bodies of standing freshwater of relatively little temperature variation— we can say that shallow bodies of standing water with little temperature variation that are not permanent are called *temporary pools*. Temporary pools periodically dry up. Some are as meager as puddles, but others are less fleeting and biologically quite complex. The most scientifically interesting of the temporary pools is the *vernal pool*.

While definitions of vernal pools proliferate, the most precise is offered by Elizabeth Colburn, a vernal pool specialist with Harvard University and the author of the definitive book *Vernal Pools: Natural History and Conservation* (2004). Colburn's definition is tied closely to the biological function of a vernal pool, specifically to the type of seasonal habitat this temporary pool can provide. According to Colburn, a temporary pool classifies as a vernal pool if it meets five factors. Vernal pools must (1) be *small and shallow*, (2) be *isolated from other bodies of water and other wetlands*, (3) *fill seasonally and at least in some years*

Smallmouth salamander larvae (one shown here) are closely related to Jefferson salamanders. Both species are common breeders in Ohio ponds and vernal pools.

dry completely, (4) *be situated in or near a woodland,* and (5) *sustain a specific biological community that lacks fish but includes wood frogs and various mole salamanders.*

First, vernal pools are small and shallow. Relative to lakes and ponds, vernal pools contain small volumes of water. Evaporation, seepage, and transpiration from vegetation in and around the pool are sufficient to draw down water levels, sometimes completely.

Second, the shallow waters of vernal pools are hydrologically isolated. That is, vernal pools lie in basins that fill from precipitation, watershed runoff, and groundwater seepage. They are not fed by streams or inundated by neighboring lakes, although some vernal pools do lie in flood plains and from time to time fill when streams or rivers overfill their banks. Nonetheless, these vernal pools are isolated for the vast majority of the time. While ponds typically have streams that drain their permanent waters, vernal pools lie in highly impermeable basins, which fill and dry without any regular outflow. This isolation is important in defining a temporary pool, for it is the very lack of steady inflow that allows vernal pools to dry up—something that helps assure that fish do not reach the pool.

Spotted salamander egg masses are vulnerable to many predators. Red-spotted newts consume large numbers of amphibian eggs, including those of mole salamanders.

The high-pitched trilling of spring peepers (*Hyla crucifer*) can be heard as early as February. The X on its back is a good field mark. The species name, *crucifer*, means "cross-bearing" in Latin.

Third, vernal pools must eventually dry out. They fill with water during wet periods, and they lose their water partially and sometimes completely, typically in late summer. Most vernal pools act as their name implies, filling in the spring, but some actually begin collecting water in the wet weather of fall. Volume may be added throughout the winter, and then the high water mark is attained in mid- to late spring. Other pools see no appreciable filling in the fall but fill throughout the winter and spring. Drying times also vary. Some pools dry up in as little as about three months, while others may retain some water throughout several seasonal cycles, only experiencing complete drawdown during droughts. In some pools, complete drawdowns may be separated by a handful of years. Generally, vernal pools will dry up completely at least every three to five years.

Fourth, vernal pools are situated in or near forests. While temporary pools may form away from woodlands, these pools differ in their hydrology and biological communities. The sun's effects are pronounced in nonwoodland pools, changing the pool's food web, increasing water temperature, and greatly affecting evaporation. Moreover, the species of animals using this type of pool are quite different, since woodland animals that migrate to the pools are absent. While biologically interesting

The translucent stabilizers on this recently hatched four-toed salamander larva help keep it upright until limbs develop.

themselves, these open temporary pools—such as can be found in coastal dunes along Lake Erie—are not discussed in this book.

Finally, vernal pools are defined by the specific biological community associated with them, at least part of which is tied directly to forests. Thus, defining a vernal pool includes understanding the *obligate species* of vernal pools. Sometimes referred to as *indicator species*, these species of vernal pools are those organisms that must be present for a pool to be considered a vernal pool. The survival of these specific species depends upon periodic drying—or at least these obligate species

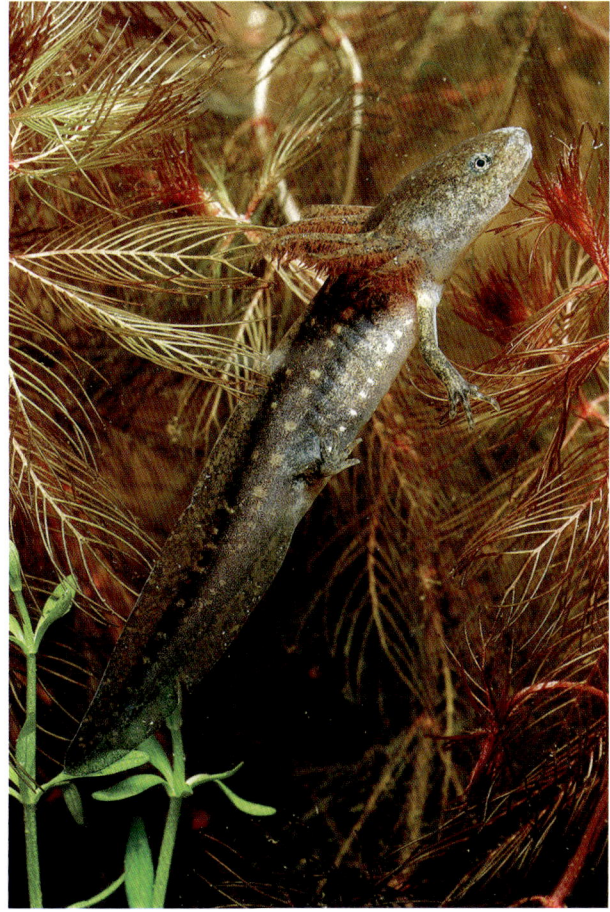

Finally, the plants that live in and around a vernal pool can affect the hydrology. Water is lost from the pool as trees and shrubs surrounding the pool, as well as the macrophytes within the pool itself, take in water from the pool and release it as vapor. This process, by which plants absorb water through their roots and other tissues and then release it into the air as water vapor through their leaves, is known as *transpiration*. Inverse relationships exist here: on the one hand, the greater the forest cover, the less the evaporation and the greater the transpiration from trees; on the other hand, the less canopy that covers a vernal pool, the more the evaporation (due to the sun's energy) but the less the transpiration from macrophytes in and around the pool.

Left: Large numbers of American toad tadpoles gather in shallow vernal pools.

Right: Marbled salamanders breed in autumn, allowing their smaller larvae to have a competitive edge over spring-hatching salamanders, which they sometimes eat.

Water Quality

Water quality is another factor to consider for vernal pools. Water quality takes into consideration physical, chemical, and biological properties of water. One physical quality important to vernal pools is temperature. On the one hand, the waters of vernal pools must stay cool enough to allow growth and reproduction of obligate species; on the other hand, many animals are positively affected by warming waters. Thermal changes trigger the hatching of eggs or the metamorphosing of young individuals into adults. The tree canopy is important in keeping pool waters from getting too hot. Of course, Ohio's moderate climate generally leads to seasonally ideal water temperatures in many vernal pools.

Another physical quality that affects vernal pools is siltation. Exposed soil and heavy precipitation can bring water that is laden with soil particles to vernal pools. The effects can be twofold. First, pools may begin to rapidly fill in, becoming too shallow to stay cool or hold water for long enough periods. Or water may become cloudy, reducing light transmission and therefore affecting both photosynthesis and photolysis, the physical breaking down of organic compounds into inorganic ones by light waves. If inorganic components are lacking in a vernal pool, primary production can flag and the pool's vitality can suffer.

Chemical qualities of water include the acidity of water as well as contaminants. Natural rainfall has a slightly acidic nature, but increases in industrial pollution in the last century have contributed to greater rain acidity; the western half of Ohio, however, is underlain by calcareous bedrocks. This typically results in acid *buffering*, neutralizing the deleterious effects of acid influxes on vernal pools. Chemical contamination, however, can have a much more serious effect. Increased use of agricultural fertilizers, for example, introduces elevated levels of nitrogen, which can result in *eutrophication* within pools. Eutrophication is a natural process by which bodies of water increase in nutrients, causing algae and larger plants to grow profusely. The result of increased plant growth is a depletion of dissolved oxygen in the water. In turn, the reduced levels of oxygen support fewer and fewer animals, reducing the total number of animals as well as species diversity. In extreme cases, the plants resulting from eutrophication use so much oxygen themselves that the water becomes bereft of adequate dissolved oxygen. Plants then die and decay, a process that requires oxygen, further reducing levels of O_2. Oxygen depletion can become a vicious cycle of death and decay, leading to highly unproductive, oxygen-depleted waters. While eutrophication is a natural process of succession in both ponds and pools, it can be greatly accelerated by human activities, such as the introduction of fertilizer to watersheds. Besides fertilizers, other factors affecting water quality include other agricultural chemicals, industrial pollution, including the release of heavy metals like copper and mercury, and the introduction of animal and human waste into a pool's watershed.

Full of eggs, this female wood frog is ready to mate. Wood frog egg masses sometimes cover large areas of vernal pools.

As you can see, what surrounds a pool has a great impact on water quality and the vitality of the pool's inhabitants. Undisturbed forests and other naturally occurring environments around a vernal pool will tend to provide homes for a variety of animals and do little to adversely affect water quality. Some forms of human intervention, however—whether agricultural disturbances, industrial construction, pollution, changes in the watershed due to roads and building associated with urban sprawl, the wintertime spreading of road salt, or likely global warming—can have far-reaching effects on the variables that combine to create a vernal pool.

The seasonal cycle of filling and drying makes for a unique ecosystem not suitable for all organisms. In the case of plants, no specialized species inhabit vernal pools. Aquatic or wetland plants of many types find their way to vernal pools, often by hanging on to animals or having been deposited as seeds in waste excretions. Plants such as cattails, buttonbush, pond lilies, and many forms of algae live in vernal pools.

Special adaptations of a number of animals have made vernal pools the unique home of a number of obligate species. The ways that animals have evolved to live in vernal pools ranges from various methods of surviving drought to using the pools only during the wet periods. Further, animal species have developed ways of sensing factors such as water temperature and drying in order to trigger periods of growth or rest. On cue, many species hatch, begin growth, or move from the pool. The many animals that cannot tolerate increasing temperatures and drying avoid vernal pools in favor of more permanent waters. Many predatory animals, including fish and frogs, such as green frogs and bullfrogs, live elsewhere.

The number of reproductive strategies employed by obligate species of vernal pools include the following:

- Production of large numbers of offspring, often smaller in size (e.g., amphibian egg masses containing over 1,000 eggs) or producing fewer but larger eggs or young (e.g., the giant water bug, whose male carries 100–150 elongated eggs upon his back until birth)
- Long life spans (e.g., mole salamanders, who can live twenty years or more)
- Complex life cycles that include a mobile adult stage, careful timing of breeding, and specialized eggs that are drought resistant (e.g., a number of vernal pool beetles)
- Short life spans along with rapid growth and breeding that culminates in large numbers of drought-resistant eggs or cysts (e.g., crustaceans, aquatic insects, and fairy shrimp)
- Continuous breeding (i.e., multiple generations in the same year) along with drought-resistant eggs, cysts, or aestivating stages during pool drying (e.g., certain snails and crustaceans)[1]

Animals in and around vernal pools can be classified with respect to their breeding and permanence. Three categories of animals include migratory breeders, nonbreeding migrants, and permanent residents. Wood frogs and mole salamanders are good examples of migratory breeders. These amphibians leave the pool after breeding or larval growth and

TABLE 2.1. TYPICAL ANIMALS IN OHIO VERNAL POOLS

Salamanders
- red-spotted newt
- spotted salamander*
- tiger salamander*
- small-mouthed salamander*
- marbled salamander*
- Jefferson salamander*
- four-toed salamander*

Frogs and toads
- American toad
- Fowler's toad
- bullfrog
- northern green frog
- northern leopard frog
- wood frog*

western chorus frog
- spring peeper
- gray treefrog
- Blanchard's cricket frog
- mountain chorus frog

Turtles
- spotted turtle
- wood turtle
- snapping turtle
- midland painted turtle
- eastern box turtle

Snakes
- northern water snake
- eastern garter snake
- eastern ribbon snake

Invertebrates
- caddisfly larvae
- predaceous diving beetle
- crawling water beetle
- water scavenger beetle
- whirligig beetle
- damselfly
- dragonfly
- backswimmer
- water boatman
- water scorpion
- giant water bug
- fairy shrimp
- amphibious snail
- fingernail clam
- leech

*Obligate species are marked with an asterisk.
Adapted from the Ohio Vernal Pool Partnership's pool monitoring form.

live in the surrounding woodlands. Reptiles (such as turtles and snakes) as well as birds and mammals are examples of nonbreeding migrants, animals who breed elsewhere but utilize the environs of vernal pools for feeding. Many aquatic insects, fingernail clams, flatworms, fairy shrimp, and good numbers of crustaceans are permanent residents of vernal pool ecosystems. Table 2.1 shows some of the more common animals—obligate and facultative—inhabiting vernal pools around Ohio.

Nutrient Cycles

Just as ponds may be placed along a continuum from autochthonous production (capturing sun-derived energy within) to allochthonous production (depending on organisms outside to harness the sun's energy), so too may vernal pools; vernal pools, however, depend heavily upon outside production. That is, the nutrient cycle in vernal pools relies to a large extent on the input of organic matter from outside the pool basin. The greatest amount of energy introduced into most vernal pools is through detritus, either in the form of leaves falling from the forest canopy or through organism tissues and waste being carried into the pool in precipitation runoff.

Of course, some primary production occurs in vernal pools. In pools not under dense forest canopies, aquatic plants will produce energy through photosynthesis. The level of this primary production is often significantly less than in ponds and other larger bodies of water.

Once organic matter from dying pool macrophytes, the forest canopy, and other organisms in and around the water enters the pool as detritus, bacteria and fungi begin to feed upon it. These decomposers derive their energy from materials such as leaves and animal tissues and waste. In the process, these microorganisms also convert complex organic compounds into simpler inorganic ones, which they release into the pool. In addition, the energy derived from the detritus by decomposers is stored within the decomposers themselves, and they then may be eaten by other consumers. The energy is passed along. Various zooplankton, such as daphnia, copepods, and rotifers, feed upon the decomposer organisms.

In some cases, the decomposers are not necessary in the energy transfer process in vernal pools. *Shredders* like the water boatman, amphibious snails, caddisfly larvae, and some crustaceans may also eat fallen leaves, consuming them in tiny, torn pieces. Most of the transfer of energy from detritus to other animals in the food chain, however, occurs through bacteria and fungi.

Predators continue the food web, eating both decomposers and those organisms that feed directly upon plants. Energy is transferred from smaller organisms to larger ones. For example, young salamander larvae, such as spotted or Jefferson, eat various zooplankton. Other salamander larvae, such as the fall-hatching marbled salamander, can eat their spring-hatching cousins. Garter snakes may find their way to vernal pools and begin eating various amphibians. And snakes may end up as food for hunting raptors. The same 10 percent ratio found in ponds applies to vernal pools. That is, only about one-tenth of the energy found in one level of the food chain is passed on to the next higher level.

The drying process of vernal pools must be considered in the nutrient cycle. As the pool dries, some animals leave, some become dormant as they aestivate or rest in drought-resistant eggs or cysts, and others die. As organisms die with pool drying, the new detritus that their tissues add to vernal pools is decomposed less and less by aquatic bacteria and fungi, and more and more by terrestrial organisms. It is significant that as the pool dries and terrestrial fungi decompose detritus, the nutrients they leave behind are different in nature, showing, for example, higher levels of protein.[2] The added nutritional value, therefore, helps increase the productivity of vernal pools. Lastly, animals not found in the pool during flooding may be found wandering through the muddy basin, leaving behind additional waste as well as seeds for future plants.

Notes

1. Colburn, Tyning, and Leahy, *Pond Watchers Guide*, 73, 75.
2. Ibid.
3. Ibid.

Below the Surface

Non-Insect Invertebrates

S itting by a pond on a summer's afternoon, you cannot help but see invertebrates: darting dragonflies hunting among cattails, tandem damselflies dropping eggs into the water, and agile water striders skating along the pond's surface, dashing among whirligig beetles and water boatmen. Kneeling along the shoreline, you might encounter crayfish flipping their tails to shoot backward and hide, snails climbing stalks of emergent vegetation, and spiders spinning webs and waiting for prey. Sticking your hand into the mud, silt, and rocks of the pond's bottom, you find a handful of fingernail clams, leeches inching along in search of parasitic hosts, wriggling aquatic earthworms, and a tightly closed mussel shell or two. Late in the summer, you might even see a bloom of hundreds of coin-sized freshwater jellyfish, rising to the surface of clear-water ponds.

The most plentiful of all types of living organisms, invertebrates—animals without backbones—account for over two-thirds of all known species on earth and over 95 percent of all animals. While the majority of invertebrates are terrestrial, a good number of invertebrates live in water for at least part of their life cycles. In this chapter and in the next are descriptions of a number of common invertebrates found in our state's ponds and vernal pools. Because invertebrates comprise such a large group of organisms, this chapter deals with non-insect invertebrates found in ponds and pools, and the following chapter (Chapter 4) discusses the diverse and populous order of insects.

The sequencing of invertebrates that follows generally moves evolutionarily, starting with the most primitive organisms and building up to the most advanced. The animals included here and in Chapter 4 are common in lentic waters, although many of these organisms can also be found in lotic waters such as rivers and streams. A few invertebrates

Opposite: Artificial light helps reveal the iridescent colors of fairy shrimp. These crustaceans swim on their backs, propelling themselves with their many legs.

that are not truly aquatic are listed here because they are frequent inhabitants in and around ponds and vernal pools.

Lastly, the invertebrates covered below are larger species. Distinguishing *microinvertebrates* from *macroinvertebrates* is not a simple task; rather, size distinctions are along a continuum. Smaller invertebrates that can sometimes be seen without magnification, like copepods and rotifers, have not been included, although excellent guides to studying such microorganisms are listed in the helpful resources section.

Flatworms, Gordian Worms, Aquatic Earthworms, and Leeches

The majority of flatworms (phylum *Platyhelminthes*, class *Turbellaria*) are microscopic and not covered in this book, but a smaller group of these free-living organisms, called *triclads* or *planarians* (order *Tricladida*), can grow to over one inch long. Planarians are often one-quarter to three-quarter inches long and dark gray, brown, or black. As their name suggests, flatworms have flat bodies, on top of which two "cross-eyed" spots can be seen. Young flatworms look much like older individuals since triclads do not go through any noticeable metamorphosis. Unless you turn over rocks in a pond or pool, you might not find flatworms since they are *photonegative*, meaning that they move away from light during the day.

By excreting a mucus from their undersides, flatworms can move their cilia and glide from one location to another. This locomotion allows some flatworms to find prey. Since triclads have no teeth, they simply attach to an organism, insert their feeding tubes, and use suction to remove fluids from host organisms. All flatworms scavenge, gathering dead animal matter along with other detritus, bacteria, algae, and protozoans. Breathing is accomplished by absorbing dissolved oxygen across their entire bodies. All triclads are *hermaphroditic*, but typically they mate with other individuals. Interestingly, if a flatworm is divided, most of the pieces will grow into complete flatworms.

Gordian worms (phylum *Nematomorpha*) are so named because of their tendency to congregate and become entangled with each other, forming a "Gordian" knot. They are also referred to as horsehair worms because their tangled masses look like clumps of equine hair. Each adult is less than one-tenth of an inch in diameter but can range in length from a few inches to over two feet, so a cluster of Gordian worms can be quite a sight. These worms can be a variety of colors, including yellowish, gray, tan, dark brown, and blackish, and some have spots or blotches.

Larvae of Gordian worms often encyst themselves on vegetation and

Almost seventy species of leeches are found in North America. All require microscopic identification.

are eaten by a variety of organisms, including crickets, grasshoppers, beetles, and other insects. Once consumed, parasitic larvae feed internally on their hosts. If the host falls into water when the worm has reached its adult stage, the Gordian worm enters the water and becomes free-living, capable of limited locomotion through basic wriggling of the body. Mating typically occurs in the spring, so adults that emerge later may hibernate along the water's edge until the following spring.

Aquatic earthworms and leeches (phylum *Annelida*) are referred to as segmented worms. These annelids are muscular, divided into discernible segments (less so for leeches), and soft-skinned. They have no head, mouthparts, antennae, or appendages. Not undergoing any noticeable metamorphosis, young aquatic earthworms and leeches appear like smaller versions of adults. The name "annelids" derives from the Latin word *annellus*, which means "little ring" (i.e., their bodies appear to be comprised of a series of rings).

Most people are quite familiar with terrestrial earthworms, but few realize that the most abundant animals in the bottoms of ponds are these aquatic counterparts. From the shallows of a pond to its deepest parts, aquatic earthworms (class *Oligochaeta*) live in two to four inches of sediment along the bottom. Adults range in length from under an inch up to about six inches, and they, like the flatworms, are hermaphroditic. They can reproduce either sexually (usually with another individual of their species) or asexually by dividing.

Aquatic earthworms are crucially important to pond life for two reasons. First, as they burrow in the sediment, they increase the dissolved oxygen within the deposits on the bottom. This creates *aerobic* conditions suitable for a diversity of organisms. Second, due to the large

NON-INSECT INVERTEBRATES   39

populations of these annelids in most ponds, they provide an important food source for other invertebrates and fish.

Leeches (class *Hirudinea*), unlike earthworms, are primarily aquatic, and most live in freshwater. Some of these annelids are less than one-quarter inch long, while others can exceed one foot when fully extended. The tops of leeches are often patterned in bright colors such as red, yellow, orange, and green. All leeches have exactly thirty-four segments, with the first few containing small eyespots on top. The two suckers, one on the front and one on the rear, are located on the bottom and are used in locomotion. Leeches move along like inchworms. The front sucker of leeches is used to attach to hosts. Some suck fluids from the small prey they kill; others are bloodsuckers.

Those leeches that suck blood also secrete an anticoagulant when they bite, causing the notorious ongoing flow of blood. Blood-sucking leeches often attach temporarily to frogs, turtles, and fish. Some prefer warm-blooded hosts, such as waterfowl and mammals (including the occasional human). Leeches have been used medicinally throughout history, practitioners believing that these annelids were removing "bad" fluids. While this practice mostly disappeared in the West during the late twentieth century, leeches are used today to remove blood around tissue transplants and reattachments. While many may consider leeches a nuisance, in fact, they, like their close relatives, aquatic earthworms, are an important food source. Leeches are eaten by fish, newts, salamanders, snakes, birds, and the larvae of insects such as dragonflies, true bugs, dobsonflies, beetles, and caddisflies.

Snails, Clams, and Mussels

Mollusks (phylum *Mollusca*), most recognizable by their hard shells, are divided into two distinct groups. The first of these are the snails (class *Gastropoda*). Gastropods tend to be terrestrial, with only a small portion living aquatically. All snails evolved from ancient gilled forms, but some of today's aquatic species evolved on land and then returned to the water. These are the lunged snails, which are sometimes called pouch snails. In fact, the "lung" in these pouch snails is actually a cavity that fills with air (or sometimes water), the lining of which absorbs oxygen (or dissolved oxygen). Because pouch snails depend less on constant water movement over their gills, some can hibernate in mud during freezes or aestivate during summer drying. Yet, while lunged snails tend to live for a year or less, most gilled snails live from two to five years.

Lymnaeid snails often feed on the pond's surface film.

All gastropods have a spiral-shaped shell that can vary in size from a fraction of an inch to about three inches. They tend to thrive in hard water, typically water that flows through limestone or dolomite, because the calcium carbonate in the water is used in building their shells. Snails with elongated, spiraled shells either have whorls that begin on the right (*dextral*) or the left (*sinistral*) when the opening is down and facing you. This distinction can aid in identification.

Snails are important for ponds both as consumers and as a food source. As snails glide on the mucus that they secrete, they extend a tonguelike appendage called a *radula*. The radula is covered with sharp hooks. The radula scrapes algae from plants and drags it, along with detritus and occasional small invertebrates, back into the gastropod's mouth. The snail's consumption of algae is important for both algae and aquatic plants. Similar to mowing a yard, the removal of a layer of algae eliminates older tissue and stimulates new growth. For the plants on which the algae grows, the snail's feeding reduces the shading effect of the algae and, therefore, increases photosynthetic efficiency. Snails are eaten by fish, amphibians, and waterfowl, all of which crush the shells, and by leeches, crayfish, water beetle larvae, true bugs, true flies, and dragonfly larvae, all of which enter the shell to feed.

Pond snails remove algae from vegetation. This planorbid is one of the state's abundant species.

Clams and mussels (class *Pelecypoda*) comprise the second group of mollusks. Most notably, clams and mussels can be distinguished from snails in that they have two shells joined by a ligament called a hinge. Clams and mussels move water through their shells using two *siphons*, one that pulls the water in and another that pushes it out; hence clams and mussels are also called *bivalves*. The circulated water is moved over gills, where oxygen is absorbed and where food, namely microscopic algae, bacteria, and detritus, is moved by cilia to the mouth. While a bivalve's siphon is extended into the water, its foot pushes against the bottom for a limited form of locomotion.

The fingernail clam (family Sphaeriidae spp.) is a common mollusk found in vernal pools as well as ponds. Fingernail clams are aptly named, as their thin and fragile shells grow to be about the size of fingernails, roughly half an inch long in adults. They are also called pill clams. These tiny clams can survive dry periods by burrowing into mud, allowing them to take advantage of the smaller number of predators in temporary aquatic environments. Female fingernail clams incubate eggs within chambers in their gills, and young are raised there. When the female releases her young, the juveniles look like miniature adults.

The zebra mussel (Dreissena polymorpha) is a common but problematic mollusk in Ohio's waters, and it reproduces quite differently. It releases from 10,000 to one million eggs per year into the water, and these may freely float and travel for up to a month, sometimes being carried to other bodies of water. The zebra mussel, therefore, can spread widely. This nonindigenous mussel can increase its population quickly and upset an entire aquatic ecosystem. The highly efficient filter feeding of masses of zebra mussels can quickly remove suspended material from water, reducing food sources for a variety of invertebrates.

While nonnative mollusks can be a problem for ponds and vernal pools, bivalves offer many benefits. Clams and mussels are eaten by birds, turtles, frogs, fish, crayfish, raccoons, and muskrats, and the filter feeding of bivalves helps cycle detritus and stirs up bottom material, creating suspended particles upon which other organisms can feed.

Water Mites and Spiders

Eighty-five percent of all invertebrates are in the phylum *Arthropoda*. As a group they make up over 80 percent of all known animals. Arthropods include arachnids, crustaceans, and insects, and they generally have segmented bodies divided into three regions (typically head, thorax, and abdomen), a hard exoskeleton made up of *chitin*, and paired, segmented legs, from which their name is derived (*arthropod* is Greek for "joint-footed").

While many arachnids (class *Arachnida*) notoriously live on land—spiders, scorpions, mites, and ticks—some are also aquatic. The aquatic arachnids are mites and spiders. Water mites (order *Acariformes*) are quite small, a little less than one-eighth of an inch as adults, with four pairs of legs, a combined head and thorax, called a *cephalothorax*, and an abdomen. Most often, they are noticed as red-colored, spider-looking animals moving rapidly on the surface of the water (although there

Snapping turtles battle for dominance. Biting matches can last more than one hour.

are many other colors of water mites). Adults have two fanglike projections called *chelicerae*, which are often used to pierce their prey; some adults, however, are herbivorous or detrivorous, and some are parasitic. Water mites undergo metamorphosis from their larval forms, which have only three pairs of legs and are all parasitic. Their complex life cycle includes the following stages: egg, prelarva, larva, protonymph, deutonymph, tritonymph, and adult. Scientists are still researching the complexity of water mite metamorphosis.

Of all spiders (order *Araneae*), only a few can live under water for any period of time, but some spiders are semiaquatic, meaning that they live around water or sometimes move across its surface in search of food. The six-spotted fishing spider (*Dolomedes tritonis*) is common around ponds and vernal pools. Its approximately one-inch-long body is hairy and patterned in tan, brown, and black. During hunting, six-spotted fishing spiders skate across the water—much like water striders—mostly in pursuit of, ironically, another invertebrate that walks on water: water striders. They can also reach underwater to catch tadpoles and even small fish. Remarkably, fishing spiders have three ways of moving on water: sailing with the wind, rowing with their middle two pairs of legs, and catapulting themselves forward with their back legs, much like a basilisk lizard.[1]

Crayfish and Shrimp

The second group of arthropods are crustaceans (subphylum *Crustacea*). They do not undergo metamorphosis. Crustaceans have many body segments with segmented appendages, horizontally moving mouthparts, and antennae, which are often called feelers. While crustaceans are diverse, the majority of them are marine species. Of the remaining freshwater types, a good number are microscopic, floating freely in the water. The larger bottom-dwelling crustaceans belong to the class *Malacostraca*, of which the most easily recognized are the decapods (order *Decapoda*), including crayfish and shrimp.

Crayfish look like miniature lobsters, which are also decapods. They tend to be brown-green, but some are blackish, red, orange, or occasionally blue. Crayfish have the ability to change color to match their surroundings. Sizes range from less than an inch to about six inches.

Crayfish can be found in large numbers in Ohio ponds. They play an important role in consuming large amounts of plant material.

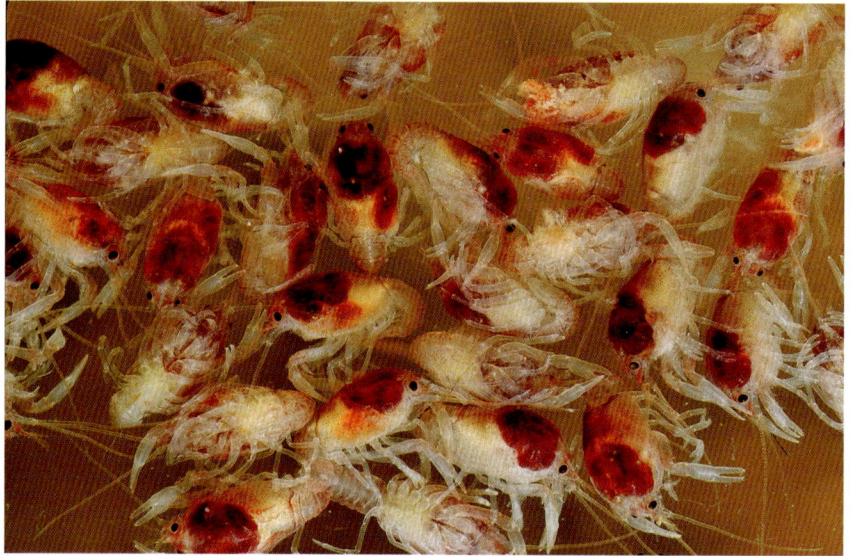

A mother crayfish releases her young from her pleopods just after hatching.

Most crayfish are photonegative, hunting at night. If disturbed, crayfish quickly dart backward by flipping their tails forward. Burrowing crayfish may be found surrounding ponds, where the crayfish dig down to the water level in order to keep their gills wet for breathing. "Crawdads" or "crawfish," as they are sometimes called, are omnivorous. Pond crayfish (*Procambarus* species) primarily eat detritus, namely decaying plant material, but they also scrape algae off aquatic vegetation, graze on aquatic plants, often cutting them off at the bottom, and prey on animals such as snails, aquatic insects, scuds, fish, and fish eggs.

A female crayfish attaches its hundred or more eggs to the underside of her abdomen and then moves her paddlelike *swimmerets* to keep water, rich in dissolved oxygen, passing over them. Crayfish do not go through metamorphosis; instead, they shed their skins two dozen or more times in their lives (which typically are two to four years long). Each time they molt or shed their skins, their bodies grow larger. Right after molting, crayfish are vulnerable until their exoskeletons harden. Besides defending themselves with their large claws and fleeing with their quick propulsion backward, crayfish legs break off easily to escape from predators. New legs, although smaller, regenerate with molting.

Freshwater shrimp (family Palaemonidae) look similar to marine shrimp. Coloration ranges from light gray to cream, and their bodies are translucent or nearly transparent. Like crayfish, shrimp are also able to change colors for camouflaged hiding. Adults generally range in size from just over an inch to almost seven inches; some, however, can grow to almost eleven inches. Larger shrimp are sometimes called prawns.

Shrimp generally live in dense aquatic vegetation, and they move by means of paddling their many segmented appendages. Like crayfish, they can shoot backward, in this case by straightening their bodies. Shrimp tend to be scrapers, removing algae from aquatic vegetation, although some also engulf young aquatic insects. Females protect their brood of eggs, but young are free-living at birth. Shrimp tend to only live one year. They are a good food source for fish.

Another type of crustacean found in vernal pools is a small creature akin to the novelty pet called "sea monkeys," to which you add water and watch the tiny animals grow. Fairy shrimp (order *Branchiopoda*) do not lay eggs. Instead, females deposit cysts—egglike structures containing fully developed larvae waiting to hatch. These cysts must go through a period of *diapause* or rest during pool drawdown. They become dry and, after rewetting, fairy shrimp emerge. Such a life cycle is matched perfectly with the filling and drying of vernal pools. About thirty hours after getting wet, the larvae emerge, typically early enough in the spring to avoid strong predation by salamanders, frogs, and aquatic insects. Fairy shrimp are indicator species in vernal pools.

Fairy shrimp are one-half to one-and-one-half inches long and are typically red-orange, although their coloration can vary greatly. If few males are present in a community, then females will lay summer cysts, which can hatch and grow into mature adults in less than a month,

A female calico crayfish carries her young attached to her swimmerets. Constant motion helps aerate the juveniles.

Above: The bladderlike traps on bladderwort contain sensitive hairs, which, when touched, trigger them to suddenly expand, capturing small swimming organisms.

Right: When aquatic sow bugs mate, the female covers the male with her body.

providing more males for winter cyst production. Finally, the survival of fairy shrimp populations is safeguarded by a low rate of hatching with inundation. One study found that only 3 percent of 1,000 cysts hatched upon first flooding, and some pools actually go a year or more without adults and then are populated again.[2]

Aquatic Sow Bugs and Side Swimmers

Two other types of crustaceans, while fairly small, also are seen with the naked eye: aquatic sow bugs (order *Isopoda*) and side swimmers (order *Amphipoda*). Sow bugs range from one-quarter to three-quarter inches long and are gray with shades of black and brown and occasional reddish or yellowish tones. *Isopods* move by crawling slowly. They are omnivorous, feeding on plant and animal detritus, aquatic plants, and live or injured animals. They have flat bodies and seven pairs of legs. A female aquatic sow bug holds her offspring in a pouch, called a *marsupium*, on the underside of her thorax. Like crayfish, the mother keeps oxygenated water moving over her young.

About the same size as aquatic sow bugs, sideswimmers—also called scuds or amphipods—are colored something like shrimp: light gray or light brown and translucent. Some have bright coloration, including green, blue, purple, and red. These flealike crustaceans have flat bodies and tend to live in shallow waters without fish, although they typically are not found in temporary waters such as vernal pools. Their feeding is similar to aquatic sow bugs, but they also scrape algae, fungi, and bacteria from aquatic vegetation.

Other Invertebrates

The third and final group of arthropods are the insects, the most diverse and prolific group of living organisms on earth. They are the subject of the next chapter. Before moving on to these, it is worthwhile to mention

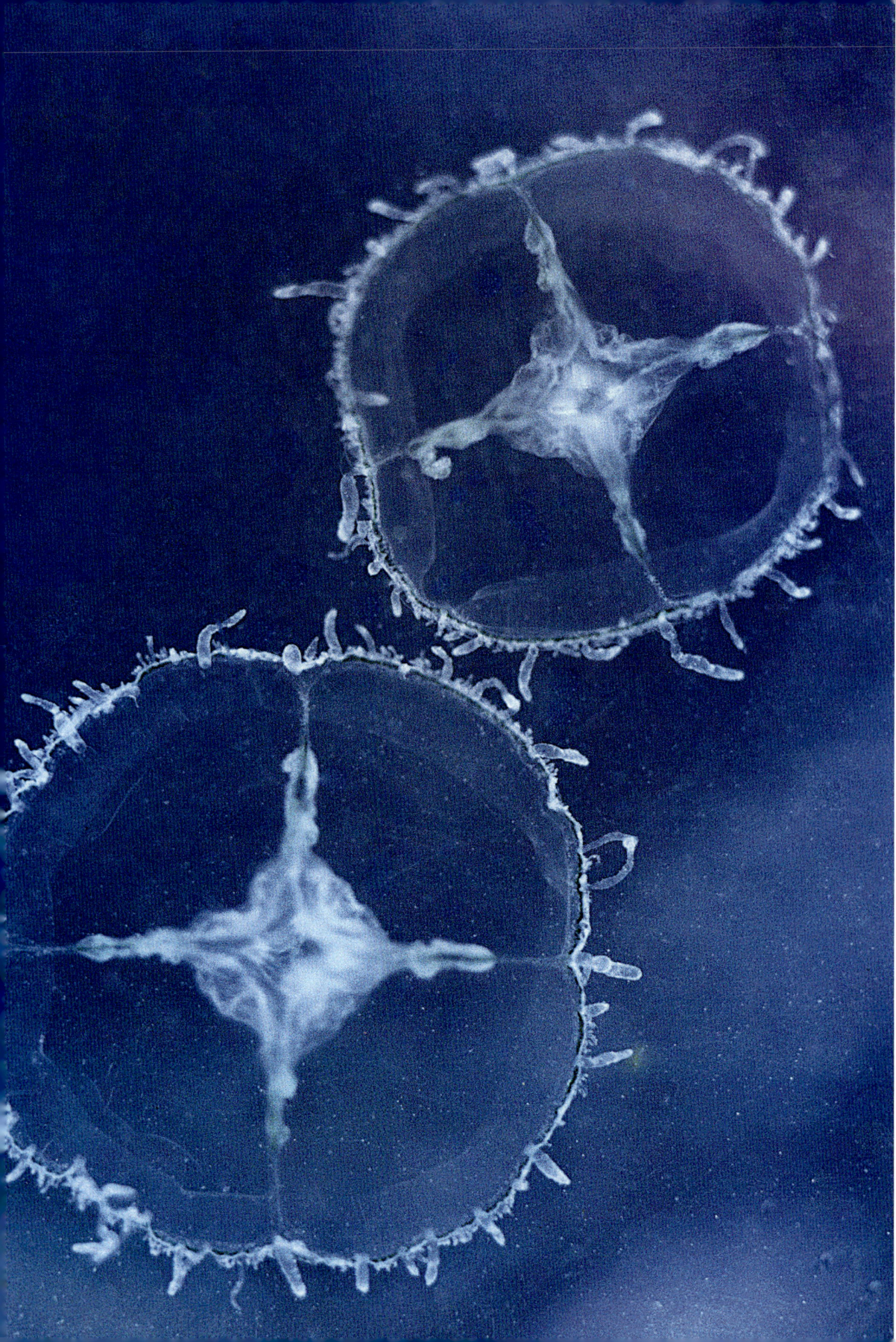

two other invertebrates that may be encountered in small bodies of water: sponges (order *Porifera*) and hydrozoans (class *Hydrozoa*), which are related to jellyfish. The vast majority of sponges are marine, but a small number are found in freshwater. Colored green, brown, gray, or yellowish and often containing algae within them, freshwater sponges are often mistaken for a form of algae. Freshwater sponges attach themselves to substrates such as rocks, aquatic vegetation, and wood.

Sponges are simple animals with no organs or true tissues, but these animals—once believed to be plants—are quite amazing. Tiny whip-like structures move water through the often finger-shaped structures, straining bacteria, phytoplankton, and detritus out of the water. A small sponge the size of this book could strain many gallons of water per day. While freshwater sponges reproduce in various hermaphroditic ways, most remarkable is their power of regeneration. If a sponge is broken up into minute pieces and placed in a dish of water, cells will join together and within several days grow a new sponge.

Among the hydrozoans, the tiny jellyfish-like hydra, typically a fraction of an inch long and usually requiring a microscope to be seen, can grow to an inch long. They vary in shape by species and have between five and twelve tentacles. Hydras attach themselves to substrates such as stones and wood and then capture their prey, often microscopic cladocerans and copepods, with their stinging tentacles. Because they can reproduce by "budding" a new individual—during which new growth seems to compensate for lost tissue—hydras are considered immortal.

Between July and October, the freshwater jellyfish can sometimes be found. Not a true jellyfish but rather a hydrozoan, much like a hydra, the coin-sized *Craspedacusta sowersbii* is translucent, with tinges of white or green. Below their bell- or umbrella-shaped bodies hang between fifty and five hundred tentacles of varying lengths; the shorter ones are used for catching zooplankton and the longer ones for locomotion. Freshwater jellyfish arrived here from Asia in the nineteenth or early twentieth century, and they inhabit all kinds of lotic and lentic environments, including ponds. Huge blooms of hundreds of medusae (the plural of *medusa*, the sexually mature stage of jellyfish), often occur just after particularly warm spells from July through October.

Notes

1. Zimmer, "Walking on Water," 30.
2. Stewart, *"Fairy Shrimp: The Ephemeral Enigma,"* February 2006, http://www. ovpp.org/CreatureFeature.html.

Opposite: *Craspedacusta* jellyfish medusa can be found in water-filled quarries and ponds. About the size of a dime, they rise to the surface with rhythmic pulsations.

Dragons, Damsels, and the Magical Six-Legged Dominion

Insects

O f all the species of living organisms on the planet, over 50 percent are insects (order *Insecta*). Among animals (kingdom *Animalia*), 70 percent are in this six-legged order, and 75 percent of all invertebrates and 85 percent of all arthropods (phylum *Arthropoda*) are insects. In short, the most diverse as well as most numerous of all macro-organisms on the planet are beetles, true bugs, dragonflies, mayflies, dobsonflies, true flies, and the like. While the majority of insects are terrestrial, they are abundant in ponds and vernal pools—indeed, *the most bountiful* in terms of numbers of species and individuals in almost all ponds and vernal pools.

Dragonflies and Damselflies

Dragonflies and damselflies (order *Odonata*) are the most celebrated of the aquatic insects, and they are for good reason. *Odonates* are numerous, large, colorful, frenetic, and fun. Observable in the late spring, summer, and early fall along the edges of ponds and pools, these impressive carnivores often have favorite hunting perches from which they dart at passing prey; they typically return to these preferred posts even after being disturbed. While their adult stages are considered only semiaquatic—living around water but not in it—dragonfly and damselfly larvae live fully below the surface and are equally tenacious in their hunting practices as their adult counterparts. Throughout the summer, odonate skins, called *exuviae*, are found on cattails and other emergent vegetation, signs of another generation metamorphosing from aquatic larvae to airborne adults. An adult that just emerged is called a *teneral*. As tenerals fill their wings with fluid and begin to allow them to harden, they cannot fly or do so weakly, making them quite vulnerable to predation.

Opposite: A green darner nymph has a hooked lower lip, which it extends at lightning speed to capture prey, such as small pickerel.

The blue dasher is one of our most common skimmers. Like most dragonfly species, they are quite colorful.

Dragonflies (suborder *Anisoptera*) are big, strong fliers. They vary in coloration and are decorated with bold patterns. Their wings are often marked with spots or *stigmas* on the tips of the wings. The order name *Odonata* fittingly is derived from the Greek word *odon*, meaning "tooth." Inspect a dragonfly head-on with a magnifying glass, and you will see some fierce mouth parts, which are used to chew prey. The aquatic larvae of dragonflies use their "teeth" to devour other insects, frog and salamander larvae, and even fish. Adults catch flying insects in their basketlike feet and are able to eat them midair. As scary as dragonflies look under a hand lens, imagine what they would have looked like in ages past. About 300 million years ago, during the late Carboniferous

Period, a dragonfly called *Meganeura* had wings as large as a medium-sized hawk, a full thirty inches across!

Dragonfly larvae are all aquatic, hatching and developing through incomplete metamorphosis (there is no pupa stage) into semiaquatic adults. Dragonfly larvae vary in size and shape, but most look similar to their adult counterparts minus wings. Their heads tend to be narrower than their thoraxes and abdomens, a trait useful in telling them apart from damselfly larvae.

Some dragonfly species reproduce in vernal pools because their eggs are able to survive periods of dryness through diapause. Adults lay eggs on stems of aquatic vegetation. As water levels drop, plants stand above the water, keeping the eggs moist as roots bring water from the mud below. When the pool dries up completely, the plants fall over, leaving dormant eggs on the bottom of the basin. When the pool begins to fill the following year, this, along with rising temperatures, triggers hatching. In most cases eggs remain dormant for three to six months, but some can diapause for close to a year before hatching.

Dragonflies are remarkable fliers due to their light weight and amazing wing control. Each of the four wings has its own independent sets of muscles, giving dragonflies remarkable maneuverability, superior, in fact, to the movement of birds. Dragonflies can fly forward, backward, hover, and change directions in mere fractions of a second. Their wings move about thirty-five times per second, and it is estimated that they can fly up to thirty-five miles per hour. Their remarkable vision—compound eyes able to see in nearly every direction—combined with their fantastic flying ability makes them fearsome predators for anything small that flies. Dragonflies catch their prey midair, grasping it between their hairy legs in what is referred to as a "basket," lifting and carrying twice their body weight.

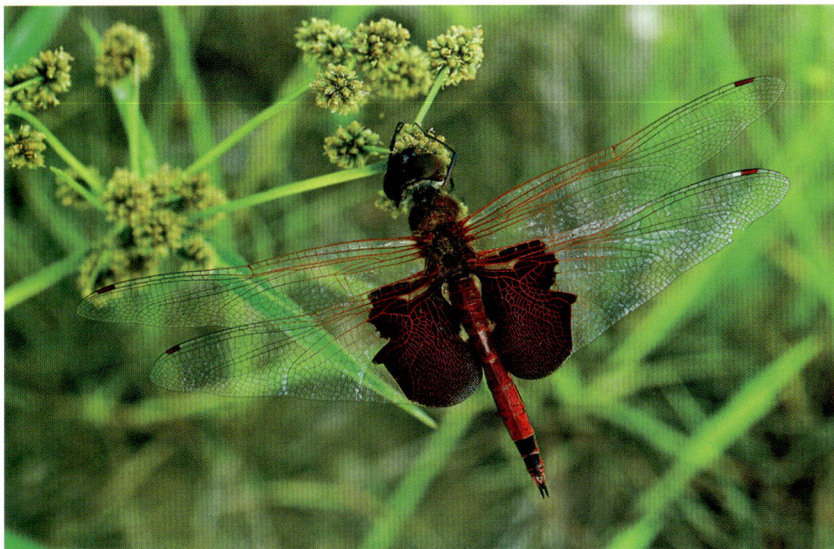

With wide wings, Carolina saddlebags can remain in the air for long periods of time.

In Ohio, the seven dragonflies families include petaltails (family *Petaluridae*), darners (family *Aeshnidae*), clubtails (family *Gomphidae*), spiketails (family *Cordulegastridae*), cruisers (family *Macromiidae*), emeralds (family *Corduliidae*), and skimmers (family *Libellulidae*), some species of which are *riverine* and some of which are *lacustrine*. The largest of these are the darners, whose long, thin abdomens look like darning needles. The common green darner (*Anax junius*) is one of about a dozen Ohio dragonflies that migrate. One generation flies south in the fall and lays eggs; offspring from these return north for summer breeding. Sometimes thousands of these large odonates can be seen swarming, ready to head

The bulbous eyes of the diminutive swamp spreadwing damselfly provide excellent vision for avoiding predators and catching prey.

south before cold fronts in late August and September, and it is not uncommon for birds, such as American kestrels and common nighthawks,
to migrate with them, feeding upon them along the way.

Damselflies (suborder *Zygoptera*) are fragile and weaker fliers. As a
general rule—to which, of course, there are exceptions—damselflies are
small, have similarly shaped front and hind wings, and close their wings
while sitting (as compared to dragonflies, which are larger and stronger
fliers, have wider hind wings at the base, and sit with spread-out wings).

Damselfly *nymphs* (the common name for odonate larvae) look somewhat similar to adult forms without wings; they have, however, three gills

that project from the rear of their abdomens. Damselfly nymphs have heads that tend to be wider than their thoraxes and abdomens, the opposite of dragonfly larvae. Like dragonflies, their eggs can rest during diapause, making damselflies common inhabitants of temporary pools.

Damselflies, like dragonflies, are often seen in *tandem*. This is when the male uses the claspers at the end of his abdomen to grasp a female on the "neck"—that is, between the head and thorax—in order to mate or lay eggs. During actual copulation, which usually occurs while perched, the male and female form the *wheel position*, with the male still holding the female and the female bringing the tip of her abdomen up to the sperm on his *hamules*, genitalia located on the second segment of the abdomen. From the side, the wheel position often forms the shape of a heart.

Over forty species of damselflies are found in Ohio. They range in color from blue, black, and green to yellow, orange, and red, and most are highly patterned on their bodies. Wings range from thin and transparent to wider and all black, many with stigmas on the wingtips. Both dragonflies and damselflies are beneficial to humans because they feed on mosquitoes, black flies, and other types of biting insects, and, despite their aggressive behavior and scary appearance, they never bite mammals, including humans.

True Bugs

While to many people "bugs" are all kinds of small invertebrates, for scientists the term refers to a specific group of insects known as true bugs (order *Hemiptera*). These insects undergo incomplete metamorphosis, with their larval stage looking much like the adult stage. For aquatic species of true bugs, both larvae and adults live in the water. They tend to live in lentic water and can be found in many of the same habitats as dragonflies and damselflies as well as aquatic beetles (see following section). While most true bugs are terrestrial, a good number of quite interesting species inhabit both ponds and vernal pools. The ability of the adults of many true bug species to fly proficiently allows them to leave temporary waters during drawdown.

True bugs vary in shapes and sizes. The larvae range from flat and oval to long and slender. They all have mouthparts that are beaklike or conical; these extend under the body when not in use. Larvae do not have full wings but instead have tiny wing pads on the thorax, where the developing wings grow. The adult forms look much like the larvae, except they have fully developed membranous wings, with *elytra* that

Opposite: Barbed ends on the lower jaw or labium can be seen in this ventral view of a green darner nymph. Sharp mouthparts help larvae devour invertebrates and even small vertebrates, including tadpoles and fish.

Some male water bug species carry eggs on their backs. Deposited by the female, the eggs are assured a fresh flow of water on the active male as well as protection from predators until hatching.

cover the hind wings. The crossing of the elytra forms a signature "X" on the back. Adults also have sucking mouthparts. All aquatic true bugs, except water boatman (family *Corixidae*), use their beaks for piercing animals. First they inject enzymes that immobilize and poison prey; then, after the enzymes have dissolved the internal tissues of their prey, they suck out the nutritious fluids.

Water bugs are fascinating creatures. Next to beetles, they are among the longest living aquatic insects. Most hatch one year and mate the next. Underwater, many communicate much like crickets or katydids, using *stridulation*, the rubbing together of body parts to create noise. While calls of *hemiptera* are beyond the range of human hearing, the sound is heard by other insects and travels well in water. True bugs do not have gills, so they depend upon drawing oxygen into their *spiracles*, just as their terrestrial relatives do. Many aquatic true bugs just rise to the surface to breathe, and some, such as water scorpions (family *Nepidae*), use a long anterior tube for breathing. They hang upside down in the water, with the breathing tube poking just above the surface.

Perhaps most fascinating is the use of air bubbles or "physical gills" to breathe. Backswimmers (family *Notonectidae*) and other true bugs—as well as some diving beetles—carry a tiny bubble of air with them underwater. To breathe, they simply pull oxygen out of this bubble. As the oxygen level of the physical gill is depleted, dissolved oxygen in the water diffuses into the bubble, maintaining a continuous level of 12 percent oxygen and 80 percent nitrogen. Carbon dioxide is released in breathing, and this diffuses out into the surrounding water. Physical gills allow insects to stay underwater indefinitely.

Backswimmers and water boatmen both "row" their bodies through the water like boats. Water boatmen use their hind legs to push through the water. They gather food by swimming along the bottom and taking in material through a conelike mouth, eating diatoms, algae, protozoa, rotifers, nematodes, and even small insects. Backswimmers move like water boatman, except that backswimmers hang upside down, just below the surface of the water. Their backs are light-colored in order to blend in with the sky, and their *venters* are dark to blend in with the bottom of the pond or pool. Though small, the backswimmer can inflict a painful bite.

Water scorpions (no relation to non-insect, terrestrial scorpions) are unique aquatic bugs. They look something like walking sticks, with long, gangly legs and equally long, slender bodies. From the hind end extends a pair of long tubes, which are used for breathing. With these tubes held just above the surface of the water, water scorpions spend much of the time with their heads held down, lying in wait for prey, which they snatch with their front legs, much like terrestrial preying mantids. Because water scorpions sit still for so long, water boatmen, backswimmers, and caddisflies often lay their eggs on them.

The most rapacious hunter among all aquatic hemiptera are giant water bugs (family *Belostomatidae*). Their shapes and colorations make them look something like small leaves suspended in the water. If giant water bugs are not able to ambush unsuspecting prey, then they use

The eastern forktail is one of our abundant odonate species. Adults can be seen from May through October.

their middle and hind legs to swim strongly in pursuit. All a giant water bug needs to do is grab a hold of its quarry, pierce its skin, and inject its poisonous enzymes. This is enough to subdue not only other insects but also other arthropods, tadpoles, salamander larvae, and even fish. In the case of some water bug species, the female lays her eggs on the male's back. He protects them and keeps fresh water flowing over them, either by swimming or, while at rest, by "rowing" his legs over them. Commonly referred to as "toe biters," giant water bugs can inflict a painful bite if disturbed by humans.

Perhaps the most interesting of the aquatic water bugs are water striders (family *Gerridae*). Able to "skate" upon the surface of the water, water striders are quick, agile hunters. Sensors on their middle and hind legs allow them to feel vibrations on the water's surface. When they sense the movement of prey, water striders move one leg forward and one backward to turn in the water, and then they "stride" across the surface of the water in pursuit of aquatic insects, which they capture with their front legs. Water striders, especially in crowded conditions, are known to cannibalize. Nonwettable hairs on the tips of their appendages allow these bugs to stay on the water's top film, not breaking the surface tension; they are able, however, to break through this physical barrier, diving under water for protection. During rainstorms they hide along the shore under vegetation.

Mayflies and Caddisflies

Mayflies (order *Ephemeroptera*) are best known in Ohio along the shores of Lake Erie, where during early summer adults emerge, swarm, and die, leaving literal sheets of slippery mayfly bodies. Swarming mayflies can be so thick that, attracted to the lights of an electrical substation in Oregon, Ohio, they short-circuited the equipment and caused a brownout in Toledo.

Ohio is home to 114 species of mayflies, and many of these inhabit smaller bodies of waters, such as ponds and vernal pools. The eggs of some species can diapause, allowing them to survive drying periods of temporary waters. When the eggs hatch, copious larvae ranging in size from less than one-quarter inch to one-and-one-quarter inches emerge, typically eating plant detritus and many becoming food for other animals. Indeed, in a good number of ecosystems, mayfly larvae are important connections in the food web, transferring the energy of primary producing plants to consumers. The larvae also eat plankton and small insects.

Mayfly larvae typically have, along their elongate bodies, seven sets of gills, which wave up and down, keeping water flowing over them. When in water rich with dissolved oxygen, the gills undulate slowly, but in oxygen-poor water, these gills move water over themselves quite rapidly. Ephemeropteran larvae shed their skin more times than any other aquatic insect, sometimes as many as forty-five times.[1]

The movement from larvae to adult involves, first, finding a safe place to emerge from the water. For some species, this occurs on the surface of the water, a location that requires a speedy emergence due to fish predation. Others crawl onto vegetation or up onto shore to emerge. The mayflies that emerge, however, are not adults yet. This *subimago* stage is brief, during which their reproductive systems develop. This stage can be as short as one or two minutes or as long as two days. Mayflies are the only insects that have fully developed wings but are not sexually mature adults.

The mayfly molts again, emerging for its extremely brief life as an adult—the stage that most people observe. This ephemeral stage is so geared for reproduction that the adults never eat. In fact, adult mayflies have no mouthparts. The males swarm and are able to be recognized by females of the same species according to the time of day, height of the swarm, number of males, and their motions. Females swoop into the mass of flying bodies and are grasped by males for mating. Females lay their eggs immediately, and both die soon afterward. While mayfly larvae can live up to two years, adults typically live only a day—some as little as ninety minutes.

Those who experience swarms of mayflies may find them a curse. The truth is that mayflies have no mouthparts and, therefore, cannot bite. While they are not quite dirty insects, cleaning up their dead bodies can

A giant water bug rests just below the water surface. A siphonlike appendage at the tip of the abdomen insures a supply of fresh air.

be a tedious chore. Nonetheless, a bucket of mayfly remains makes a great fertilizer. Most importantly, the populations of mayflies are strong indicators of healthy ecosystems. That is, the greater their numbers are, the healthier the aquatic environments are. Good adult populations indicate greater numbers of larvae, which provide food to fish, amphibians, and other larger water dwellers.

One of the aquatic insects best known to fishermen are caddisflies (order *Trichoptera*), the most diverse order of entirely aquatic insects. While many caddisflies tend to live in lotic waters, especially cooler rivers and streams, many species of this diverse group live in warmer ponds and even temporary pools. Caddisflies are closely related to moths and butterflies (order *Lepidoptera*), most notably in the ability of their larvae to make silk thread. In all but one family of caddisflies, silk is used to build a larval structure. Within ponds and vernal pools, these structures tend to be cases constructed out of sand, small stones, crushed shells, pieces of leaves or rolled leaves, twigs, and bark. The caterpillar-like larvae build elongate structures, more or less tubelike, to which they often add as they grow. Not only do these provide protection, but they also aid in respiration. Caddisfly larvae move water through their cases, increasing the flow of oxygen-rich water over their gills.

Caddisfly larvae vary in size and color (as well as their case shape and materials), and they range in size from less than one-quarter inch to almost two inches. Their diet consists of everything from detritus and plant material to zooplankton and invertebrates: members of the giant casemaker caddisflies (family *Phryganeidae*) can be found in vernal pools feeding on salamander eggs.

Trichopterans undergo complete metamorphosis, entering a pupal stage within a silken case, quite similar to the process moths undergo. Caddisflies emerge from the pupal case by cutting their way out with sharp teeth and emerge as hairy-winged adults, once again resembling their moth relatives. Adults can be quite colorful and range in size (not including their long antennae) from less than one-quarter inch to about one inch. They hold their wings over and beyond their abdomens in a slanted, rooflike position. Most adults are, like mayflies, nonfeeding, focusing their one or two remaining weeks of life on reproduction.

Beetles

While beetles (order *Coleoptera*) represent over one-third of all insects worldwide, very few of them are aquatic. Nonetheless, some of the most commonly encountered aquatic insects observed in ponds and vernal pools are water beetles. Water beetles undergo complete metamorphosis, with most living in the water during all stages of development except pupation, when they either dig burrows for their four- to six-week change from larvae into adults or build protective mud shells on the stems of nearby vegetation. Water beetles have relatively long life spans, usually about one year; some are active during the winter, and others hibernate throughout the colder months. While most people are familiar with the appearance of adult beetles—small- to medium-sized, somewhat oval bodies, and protective elytra or hardened front wings covering their foldable rear wings and abdomens—the larval stage is less familiar. Larvae tend to be more caterpillar-like, with softer, more cylindrical bodies and no wing pads. Because beetles are strong fliers, they can move from one habitat to another easily. This means that water beetles are commonly found in vernal pools, taking advantage of fishless waters and then moving elsewhere upon drawdown. The name beetle comes from the Old English *bitula*, meaning "to bite." Beetles have large, horizontally opposed mouthparts allowing for various types of feeding. A few aquatic beetles can bite humans, causing some pain.

Perhaps the best known among water beetles are whirligig beetles (family *Gyrinidae*). One-half-inch-long adult whirligig beetles can be found congregated in groups called "schools," spinning in circles on the surface of the water. Whirligig beetles have divided eyes, which allow them to see both above and below the water. It is hypothesized that the tiny waves created by their circular swimming allows them to use a form of echolocation to find prey, typically animals trapped on the water's surface. Adults also scavenge.

Many species of mosaic darners, like this female green-striped darner, emerge in late summer. Many mosaic darners look alike, making them hard to identify in the field.

Plates on whirligigs' legs act as oars, allowing them to "row" through the water. The teardrop shape of their bodies makes them excellent divers. While whirligig beetles breathe from the atmosphere when above water, they can also trap tiny bubbles of air under their elytra for breathing when diving. Toxic secretions deter predators, making some whirligig beetles smell like apples. As with many other animal species, they congregate in large groups, often in late summer, which aids in protection. Adult whirligig beetles bite, and some people experience localized allergic reactions.

The larvae of the whirligig beetles are as hard to find as the adults are easy to locate. The larvae are larger than the adults, often over one inch long, but they hide among dense aquatic vegetation. Dissolved oxygen is absorbed across their whole bodies, but whirligig larvae also breathe through filamentous gills along their abdomens, which make them look quite similar to hellgrammites (the larval stage of dobsonflies) and alderfly larvae. Larvae prey upon other invertebrates and sometimes cannibalize their own. Whirligig beetle larvae emerge from the water and build mud structures on plants and stones near the shore for pupation. Pupae are often parasitized by wasps.

Among the most voracious aquatic insects are the predaceous diving beetles (family *Dytiscidae*), which eat basically anything they can capture and overcome, including a whole host of invertebrates such as worms, leeches, crustaceans, and insects as well as tadpoles, salamanders, and even fish. In lentic habitats, dytiscids account for a large percentage of all beetles.

The larvae, called "water tigers," can reach nearly three inches in length, having cylindrical bodies with three pairs of legs, two claws on the end of each leg, a split tail, and, most interestingly, large, specialized jaws. Their jaws have channels in them: when the water tiger bites, a brown fluid is injected into its prey along the channels. This dissolves the internal tissues into a liquid, which is then sucked out of the animal through the same channels in the jaw.

Adult predaceous water beetles range in size from less than one-quarter inch to over one-and-one-half inches. They are powerful swimmers, as their hind legs are covered with hairs. When the beetles move their legs in unison, the hirsute appendages push against the water and propel the beetle forward rapidly. Air bubbles trapped under the elytra allow adults to stay underwater and breathe. They attack all kinds of prey, tearing them apart with their strong, chewing mouthparts. Adults also scavenge dead animals. They are strong fliers, allowing them to leave habitats. Attracted to light, dytiscids may be found around security lights near ponds and pools. They may live as long as three years.

Equally long-lived are some water scavenger beetles (family *Hydrophilidae*). Hydrophilids occur in the same habitats as dytiscids, and they are only exceeded in numbers of species among water beetles by predaceous diving beetles. In fact, most water beetles collected in ponds and pools belong to these two families. The larvae of the water scavenger beetles look much like dytiscid larvae; notably, water scavenger beetle larvae have teeth on their jaws, which are used to prey upon all kinds of animals, including snails, shells and all! Larvae of some species grow to two-and-one-half inches.

Adults do scavenge on dead animals, but they also prey upon small invertebrates as well as graze upon aquatic vegetation. Air bubbles are stored under the elytra for breathing. Water scavenger beetles also have nonwettable hairs on their undersides, which allow a thin film of air to adhere below, forming a *plastron*. This acts as a physical gill and makes it possible for adults to stay underwater indefinitely in water that is rich in dissolved oxygen. Adult hydrophilids often can be distinguished from dytiscids in the water by their breathing positions. Predaceous diving beetles breath through tubes at the end of their abdomens, so they are often seen tilted head-down in the water. In contrast, water scavenger beetles allow air to refill their plastron by extending one antennae above the water, so they are often seen tilted head-up in the water.

True Flies

Over half of all aquatic insects are true flies (order *Diptera*). Thus, in many ponds and vernal pools, the greatest diversity of insects are likely dipterans, yet it is adult forms that most people are familiar with, such as horse flies, deer flies, crane flies, no-see-ums, and mosquitoes. While few other orders of aquatic insects prove problematic to humans, many flies—while often beneficial ecologically—are major nuisances, cause significant economic woes, and pose some serious health risks for people. And, because they can exist in environments ranging from geyser pools and oil seeps to polar regions and highly polluted lakes, diptera are practically ubiquitous.

Dipteran larvae, while quite diverse, all have soft flesh and are elongate, most ranging from a fraction of an inch to about an inch (although some can grow quite a bit larger). Most are cylindrical and look like maggots (the common name for a number of terrestrial fly larvae). No wing pads or segmented legs occur on the thorax, although fleshy *prolegs* may be present on some species. Some larvae leave the water to pupate, and some stay underwater as pupae.

Adult forms are better known. Dipterans are so named because they have two wings used for flying instead of four (*di* meaning "double" and *ptera* meaning "wings" in Greek). The stubby rear wings are called *halteres*. Short stalks with knobs on the ends, they look nothing like wings. While all adult flies are terrestrial, most remain near their aquatic environments upon maturity. A few, especially blood-sucking females of certain families, may fly a distance to seek food necessary for egg development. Adults typically have mouthparts specialized for drinking or sponging fluids, but some have piercing tubes for sucking blood. Adults lay eggs in a variety of manners, including in the water, on the shore, and in mud. Some eggs can undergo diapause, allowing certain flies to lay eggs that can survive the drawdown of temporary pools.

Life cycles vary widely among the true flies, but generally eggs hatch in several days to a few weeks, with larvae remaining active during the winter. Larvae can live from a few weeks to two years. Pupation is short, several days to a few weeks, and adults are somewhat ephemeral, living as little as several days to a couple weeks, although some can live for months.

Horse flies and deer flies (family *Tabanidae*) are well known as nuisances in their adult stages. Strong fliers that can range far from their aquatic habitats, female horse flies and their smaller deer fly relatives acquire blood through painful bites, first cutting the skin and then sopping the blood, usually leaving painful wounds on their victims. Horse flies and deer flies are most active during hot, humid conditions. Tabanid larvae are also quite fierce, possessing two fanglike hooks for mouthparts. They tear open a hole in invertebrates and small vertebrates, insert their heads, and suck fluids from prey. Horse flies and deer flies are not vectors of diseases in North America, but they can be serious nuisances during warmer weather.

Perhaps the smallest of notorious diptera are biting midges or "no-see-ums" (family *Ceratopogonidae*). Because of their diminutive size, adult females, who also seek blood for egg production, can pass right through screens and insect netting. Despite their miniature size, no-see-ums have a rather painful bite. Biting midges also suck blood from other mammals, birds, reptiles, and amphibians as well as other insects.

Some females will even suck the blood of the male during mating. Larvae with mouthparts pointed forward are generally predators, feeding on other invertebrates and their eggs, while those with their mouths pointed down tend to feed on fine particles of detritus.

Beyond these blood feeders, there are many flies found near ponds and pools that are harmless to humans. Nonbiting midges (family *Chironomidae*) are often bountiful, most building tiny tubes along the bottom and consuming detritus. The larvae of some species contain hemoglobin in their blood and can live in oxygen-depleted waters. High concentrations of these "bloodworms" often indicate elevated levels of organic wastes or nutrients entering the ecosystem. Adults probably do

An autumn meadowhawk after emerging rests on its exoskeleton or exuvia. Most dragonflies transform at night, which protects them from predators.

not feed and are best known for their swarming, when males congregate in great numbers and grasp females who enter the swarm. In Ohio, one type of nonbiting midge that often is found swarming along Lake Erie during the summer is referred to locally as a "mucklehead."

Other nonbiting flies are common in and around lentic waters. Dance flies (family *Empididae*) are so named for their swarming just above the water, and, lucky for humans, they feed on mosquitoes emerging from their pupal skins. Many believe that crane flies (family *Tipulidae*) also eat mosquitoes. While many adult crane flies look like giant mosquitoes—with such long bodies and gangly legs that some refer to them as daddy long legs—they actually are nectar drinkers if they feed at all. Most aquatic crane fly larvae are *detrivores*.

Larvae of dixid midges (family *Dixidae*)—the adults of which are not blood feeders—often form inch-long "U" shapes in the water, moving by stretching out and then bending again into a "U" shape. Rat-tailed maggots are the larval form of surphid or "flower" flies (family *Syrphidae*). The larvae grow to about half an inch, but they have breathing tubes that exit the rear of the abdomen and can extend more than two inches to the surface of the water. Adult surphids often look like bees—except that flower flies can hover in one spot and have no stingers or other injurious body parts.

Also found near ponds and pools are other common diptera: moth flies (family *Deuterophlebiidae*), whose adults are quite hairy; soldier flies (family *Stratiomyidae*), whose larvae live in decaying vegetation, mats of algae, and muck; shore and brine flies (family *Ephydridae*), which are also capable of living in extremely harsh habitats, such as geothermal springs, tar pits, and oil seeps; phantom crane flies (family *Ptychopteridae*), whose adults almost float in the air with the aid of their gangly legs; and phantom midges (family *Chaoboridae*), whose larvae appear to float up and down in the water. Indeed, the order of flies is so diverse and so many diptera can be found around almost any type of water that the preceding list of common families serves simply as an introduction.

Note

1. Voshell, *Guide to Common Freshwater Invertebrates*.

Truthful Tales

Fish

Over 160 different species of fish can be found living in Ohio's freshwaters, but most of them are found in rivers, streams, and larger lakes, especially Lake Erie. In fact, while only about 10 percent of these species might be encountered in studying a variety of ponds across the state, it is more likely that fewer than ten species will be commonly seen. In addition, no fish are found in vernal pools. Remember, the major factor in determining a vernal pool is periodic drying, something no fish can tolerate.

Of all the animals found in Ohio's ponds, fish are the most highly managed. While some fish can be introduced naturally to ponds by birds and other animals bringing them in—usually carrying viable eggs from one body of water to another—most fish end up in ponds through human intervention. Some species are brought in by anglers, who sometimes dump fish from a previous source into a pond's water; the vast majority of fish are the result of careful stocking.

Ponds with diverse and thriving fish populations often begin with careful planning. Almost all ponds in the state have been built by humans; in a way, planning a pond is like planting a garden. First, you must determine what is able to grow in your area based on climate and soil conditions. Then, from the plants that you could grow, you select the ones that you *want* to grow. Next you determine how many of each kind to plant. From these considerations, you can then determine the size of the garden, its placement on your site, and how to acquire the vegetables you will soon plant.

Planning a pond involves first determining what fish species *could* survive based on climate, water temperature and quality, and the pond owner's level and types of interaction. Then, like planning a garden, pond managers choose what fish they want in their ponds. Unlike gardens,

Opposite: Mosquito fish, an introduced species from the South, consume large numbers of insect pests.

A breeding pumpkinseed sunfish shows off its metallic colors.

where nearly any plant can coexist with others, some species of fish live together better than others, mostly based on reproduction rates and levels of predation.

After a pond owner has determined what fish are best, then the construction of the pond is carefully planned, creating appropriate habitats for feeding and spawning. Planning for feeding is more complex than it might seem at first. Nutritional requirements begin not with the direct food source for each species of fish (e.g., bluegill for largemouth bass); rather, it starts with considering the entire food web, especially the lowest levels of detritus and fertilizers flowing into the pond as well as sustainable populations of *plankton*—both *phytoplankton* (algae) and *zooplankton*. The pond shape, especially the edges, should be designed to allow some growth of aquatic vegetation, but if more than about 10 percent of the surface area is covered with plant growth, fish populations may suffer. This means that edges typically slope quickly to depths of two-and-one-half feet. One excellent source for planning and managing a pond is the Ohio Department of Natural Resource's *Ohio Pond Management Handbook: A Guide to Managing Ponds for Fishing and Attracting Wildlife*, available online from the Division of Wildlife in electronic format. Also seek the advice of Ohio's licensed commercial fish propagators and experienced pond builders.

The fish common to Ohio's ponds are all members of the order *Perciformes* (meaning "perchlike"). The members of this large order of vertebrates are all bony fish. Their vertebrae help protect internal structures and aid in swimming. Like all fish, those found in the state's ponds are cold-blooded. They have permanent gills, which are used to draw dissolved oxygen from the water for breathing. Indeed, gills distinguish fish from all other vertebrates: fish never have to come to the surface to breathe. All pond fish have fins used for balance and swimming, and most are covered with scales, which help protect against injury and disease. Scales are not lost with age but rather grow in size, adding rings like a tree each year. Indeed, these rings can be counted to determine a fish's age. These modified skin cells are covered with mucus, which reduces friction in swimming and helps protect the fish against disease. Inside is a specialized organ called a swim bladder (also known as a gas bladder), which helps the fish remain buoyant at a specific level in the water.

Fish are guided in survival by keen senses, the least of which is vision. While their eyes are located on each side of their bodies, providing vision in nearly all directions, they are round, providing little more than close-at-hand sight. Further, their eyes have no eyelids and no irises. The latter means that changes in light intensity make it difficult for fish to see, although they can adjust their receptor cells for some variations in brightness. Rods and cones allow for vision during both the day and night.

Smell, taste, and hearing are the primary modes of perception for fish. Behind a fish's *nares* or nostrils lie highly sensitive smell receptors, which aid in locating food and recognizing predators. Fish can also taste amazingly well, sensing among flavors that are sweet, sour, salty, and bitter. Taste buds are more often external; some fish, such as catfish, have taste buds all over their bodies.

Perhaps most extraordinary are the dual methods by which fish hear. First, fish have ear structures under the skin on both sides of their heads. Sound, which passes five times faster through water than through air, strikes the fish's skin and reaches their ears. Second, fish utilize what is called an *acoustico-lateral system* for hearing. Running down each side of the fish's body are lateral lines, which are made up of tiny *neuromasts*, specialized hair cells, similar to those within the ear. These specialized hair cells perceive vibrations. The functions of the acoustico-lateral system is not fully understood, but fish may use this secondary hearing not only in food location, especially in the dark, but also in maneuvering within schools and sensing changes in the weather.

By far and away the three most common species of fish found in Ohio's ponds are bluegill, largemouth bass, and catfish. Bluegill (*Lepomis macrochirus*) are one of twenty-seven species of sunfish (family *Centrarchidae*), all native only to North America. Bluegill are the basis for most ponds' fisheries. They feed on zooplankton (which feed on phytoplankton); in turn, bluegill are the forage for a number of other species of fish. They are tall fish that are prominently ray-finned; that is, they have quite noticeable bony spines in their fins. Bluegill vary in colors, but they always have a black *opercle* or ear flap and black spot near the back of the dorsal fin.

Bluegills reproduce throughout late spring and summer, peaking in June. In one to four feet of water, males fan a nest along the bottom, where females lay 10,000 to 60,000 eggs, which the male protects until they hatch in about five days. The fry feed on zooplankton; as they mature, they feed more and more on insects. Blue gill may reach seven to ten inches and weigh up to three pounds when fully grown.

Another widely distributed and well-known fish in Ohio is the largemouth bass (*Micropterus salmoides*). Also known as black bass, they also are members of the sunfish family (i.e., they are not members of

Bluegill sunfish are our most abundant pond species. Adults may be stunted in overstocked populations.

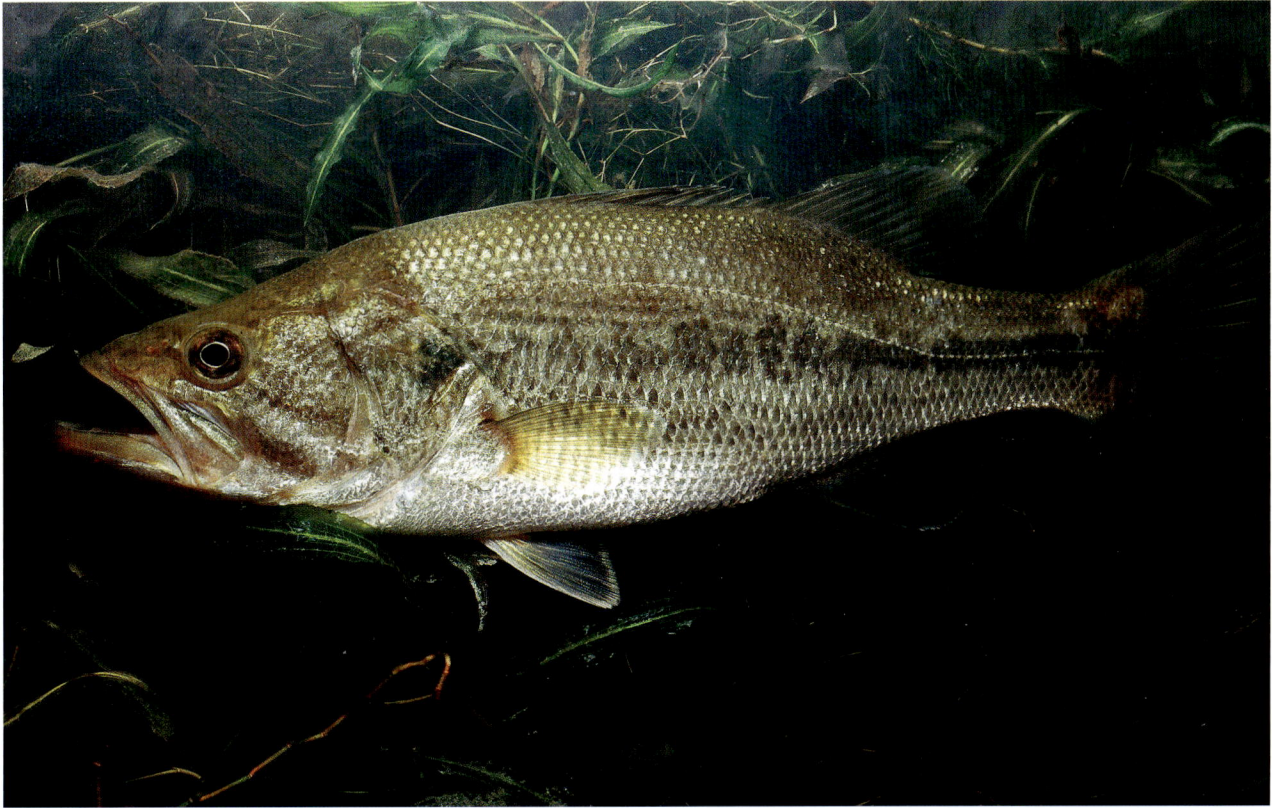

Ohio's native bass family, *Percichthyidae*). Their name derives from the fact that the largemouth bass's mouth, when closed, extends behind its eye. Largemouth bass have a single black band running along each side of their elongate bodies. These top predators in the food chain can grow to be two feet long and weigh up to ten pounds.

Spawning occurs between mid-April and mid-June. Males build nests in one to six feet of water, where one or more females deposit eggs. A female will lay anywhere from 2,000 to 20,000 eggs, depositing them in one or more nests. The male guards the eggs until hatching. Fry consume zooplankton, but after growing to larger than one inch in size, they eat freshwater shrimp, insects, amphibians, and other fish.

Three species of bullheads, members of the catfish family (*Ictaluridae*), inhabit Ohio ponds. Catfish are so named because of the barbels or "whiskers" that extend from either side of their mouths. These protrusions are covered with a high concentration of taste buds, which also cover their entire bodies. This, combined with an acute sense of smell, allows them to find food in dark or murky water. In fact, the odor receptors within the catfish's nares can sense certain compounds at a remarkable one part per 100 million parts of water. Catfish are long, each with a noticeably forked dorsal fin or tail fin. They are bluish-silver in color

Largemouth bass can reach three feet in length and are one of the pond's top predators.

The brown bullhead is one of three species of bullhead catfish found in Ohio waters.

and typically covered with dark spots. Their dorsal and pectoral fins have sharp spines, requiring careful handling. Catfish have no scales.

Catfish spawn when the water reaches seventy degrees Fahrenheit. Females lay 8,000 to 15,000 eggs in natural cavities, with both parents guarding the nest. Adult catfish are notorious bottom feeders, eating insect larvae, crayfish, snails, fish, and carrion. They provide excellent cleanup along the bottoms of ponds. Catfish are also favorite table fare. While average-sized channel catfish are about one foot in length and weigh around a pound, they can grow considerably larger, some to well over ten pounds.

Beyond these most common species, a number of other types of fish may be encountered. In addition to bluegill and largemouth bass, some other common pond-dwelling members of the sunfish family include the smallmouth bass, green sunfish, warmouth sunfish, pumpkinseed sunfish, redear sunfish, black crappie, and white crappie. Smallmouth bass (*Micropterus dolomieu*), which are yellow-green in color, can be distinguished from the largemouth by the size of the mouth, which, when closed, does not extend past the eye. Green sunfish are native to Ohio and are usually found in channel ponds.

Warmouth sunfish (*Lepomis gulosus*), a southern species, are dark brown-olive in color, striped, and mottled. They grow to about the same size as bluegill. Males protect their nests by puffing out their gill covers and opening their mouths to frighten potential predators. Pumpkinseed (*Lepomis gibbosus*) are one of the most brightly colored fish commonly encountered in Ohio ponds, with yellow and orange below and greens, blues, and other colors above. Redear sunfish (*Lepomis microlophus*) look much like bluegill except that the margin of the opercle is red or orange and no dark spot appears at the base of the dorsal fin. Growing a bit larger than bluegill, redear eat large amounts of snails using specially modified teeth that crush the mollusks' shells. To this end, redear are often called "shellcrackers."

Black crappie (*Pomoxis nigromaculatus*), typically eight to twelve inches in length when fully grown, are mottled with dark blotches that do not form a distinct pattern. They have seven or eight dorsal bony spines on the dorsal fin. Because black crappie reproduce in large numbers and are predatory, their numbers can increase to the point of overpopulation, subsequently producing stunted growth. That is, too many living in one place can result in a pond full of six- or seven-inch adults.

White crappie (*Pomoxis annularis*) are the same size as black crappie, but the white species has five to ten black vertical bands on each side. The dorsal, anal, and caudal fins are decorated with spots forming bands, and the dorsal fin has five to six spines. If white crappie occur in ponds, they arrive either through introduction by animals such as birds or through flooding from a nearby river. White crappie reproduce plentifully at such small sizes (four to four-and-one-half inches in length) that they can quickly overpopulate a pond and create stunted adults five to six inches long.

Many ponds contain smaller fish that serve as forage for the larger species. The most common of these are fathead minnows (*Pimephales promelas*). Growing to just three inches, these olive to olive-yellow minnows have slender bodies with triangular heads, lighter bellies, and a stripe down each side. Abundantly reproducing and omnivorous, fathead minnows are the primary feeder stock in many ponds. Rosy-red minnows are a strain of fathead minnows that have a light orange appearance. Fathead minnows are well known for their production of Schreckstoff, a chemical alarm signal released by injured individuals, which alerts individuals of danger.

Similar to the fathead minnow—except for a lateral band that runs all the way from the snout to the tail and a black spot next to the tail—the bluntnose minnow (*Pimephales notatus*) is one of, if not the most, common freshwater fish in the eastern United States.

Larger than fathead and bluntnose minnows, golden shiners (*Notemigonus crysoleucas*) average between two-and-one-half and seven inches in length. While their bodies are a distinctive golden olive-silver, their fins vary in color from golden brown to reddish-orange.

The mosquito fish (*Gamusia affinis*) is an introduced species whose name indicates that these small fish feed on mosquito larvae. The use of mosquito fish is controversial as their introduction to some habitats can result in pressure on other invertebrate species.

Another biological control introduced to many ponds are grass carp (*Ctenopharyngodon idella*). Native to China, grass carp consume large amounts of aquatic vegetation. Used for weed control, the silvery-green colored grass carp, also known as white amurs, can grow to thirty-five pounds or more. By five years of age, they reach maturity and begin eating less. Fine table fare, white amurs are often removed and restocked periodically. Because of their devastating effects on natural habitats, the grass carp sold by hatcheries are triploid, which means that they cannot reproduce.

Common carp (*Cyprinus carpio*), natives of Europe and relatives of common goldfish, are also stocked in ponds as biological controls for vegetation. Common carp are bronze-gold to golden yellow, have two

A grass pickerel hides in dense vegetation, waiting to snap up anything edible that happens by.

barbels on each side of the mouth, and grow to over three feet and fifty pounds. Unlike white amurs, common carp are omnivorous and reproduce. While common carp do control vegetation, they are also nest-raiders, reducing populations of other fish species. In addition, their bottom-feeding practices, rooting through the clay along a pond's basin, stirs up particles that settle on fish eggs, reducing dissolved oxygen transfer and causing devastating effects on a pond's ecology.

A central mudminnow uses its pectoral fins to climb through dense vegetation. Gulping air, it can survive in just a few inches of water.

Both Water and Land

Amphibians and Reptiles

While anglers may associate ponds with fish, for many people the most obvious and identifiable animals are various amphibians and reptiles: frogs and toads, turtles, snakes, and, perhaps, a few salamanders. Who has not enjoyed sitting by a pond on a summer day, listening to the "jug-o-rum" call of a bull frog or spotting a painted turtle basking on a log? Children squeal at the challenge of using a net to lunge at a green frog just as it springs away from the bank into the safety of deeper waters. And all stop and stare as a garter snake basks on the shore, warming in early spring sunlight, or as a water snake silently slithers "S" shapes across the duckweed-covered surface of the water, seeking a new meal of just-emerging tadpoles. Indeed, amphibians and reptiles are the universal symbols of wetlands.

Often listed as "reptiles and amphibians," evolutionarily the order should be reversed. Amphibians developed first. The name *(amphi* means "both" and *bios* means "life") indicates that amphibians live in the water and on land. Most amphibians spend their larval stages in the water and then change their body form and structures in ways that better suit them for adult life on land. *Metamorphosis*, therefore, is a key concept in the life of amphibians.

Among the amphibians, fifteen species of frogs and toads (order *Anura*) and twenty-four species of salamanders (order *Caudata*) inhabit Ohio. Of these, all but three of the frogs and toads and half of the salamanders are commonly seen in and around ponds and vernal pools. Amphibians can be differentiated from other vertebrates in that they are not covered with scales, fur, or feathers; rather, they tend to have smooth skin that is moist or slimy. Internally, their hearts have only three chambers, and they return to wet or moist locations to lay their single or multiple jelly-covered eggs.

Opposite: A male red-spotted newt grasps a female around the neck, ready to mate. The pair may remain in amplexus for over three hours, sometimes drowning the female.

Frogs and toads are the most conspicuous of Ohio's amphibians. The best known of the anurans are true frogs (family *Ranidae*). These include the American bullfrog, northern green frog, northern leopard frog, and wood frog.

The American bullfrog (*Rana catesbeiana*) has to be the most celebrated of Ohio's amphibians. Attend any frog jumping contest, and you'll likely see a host of the state's largest frog. Like all ranids, the bullfrog has strong back legs with webbed feet. The front feet are not webbed. The bullfrog grows to half a foot long and is covered with smooth, slimy skin that is green and brown on top (sometimes with brown mottling) and white with gray spots below. Bullfrogs have conspicuous tympanic membranes or ear drums right behind their eyes. In males the circle is larger than the eyes, and in females it is about the same size. Ridges of folded skin start below the eyes and extend over and around the tympanic membranes but do not extend along the sides of the frog to form dorsolateral folds, a feature found on the similar-looking northern green frog.

Bullfrogs are common and distributed across the state in all types of waters, from lakes and ponds to backyard water gardens and roadside ditches. Male bullfrogs can often be heard repeating two or three times their distinctive, deep "jug-o-rum" call, which indicates their territory. Smaller males position themselves around a larger calling male and try to intercept responsive females. Defending territories may involve wrestling matches with kicking and even biting. Bullfrogs reproduce plentifully, the female laying between 5,000 and 20,000 eggs, whose tadpole offspring require two years of growth before metamorphosis.

Maturing bullfrog tadpoles have to occasionally surface to fill their newly formed lungs with air.

A bullfrog rests beside a spatterdock blossom. Eating anything they can swallow, these "green dragons" can be detrimental to a pond's ecology.

Tadpoles feed on algae and aquatic plants. The adults feed on anything they can fit in their mouths, including insects, spiders, worms, crayfish, and snails. These "green dragons" sometimes consume vertebrates such as other frogs, fish, turtles, snakes, rodents, and even birds. Their voracious appetites can grow so large that the presence of bullfrogs can upset a pond's ecology.

Bullfrogs themselves are eaten by a variety of mammals, such as raccoons, minks, opossums, and weasels as well as water snakes, birds, especially herons, and larger fish. Bullfrogs are also taken by humans: they are the most-prized species for frog legs.

The abundant northern green frog (*Lithobates clamitans melanota*) looks a lot like the bullfrog, with green and brown coloration. Occasionally, blue individuals are found. Dots on the back are black, and the belly is white without spots. The most distinguishing feature that differentiates a green frog from a bullfrog is the pair of dorsolateral folds, two ridges of skin, one on each side, that extend from the snout backward about halfway, dividing the back of the frog from its sides. As with the bullfrog, the male has larger tympanic membranes than eyes whereas for the female they are about the same size. Mature green frogs grow to sizes a bit smaller than bullfrogs, sometimes exceeding four inches in length.

Northern green frogs are common across the state. Their calls are unmistakable. Males expand a pair of vocal sacs, making their throats noticeably swell and producing a sound akin to the plucking of a loose banjo string, "brongggggg." During mating season, males have bright yellow throats and swollen thumbs and forelimbs, which aid in amplexus. As is the case with bullfrogs, females select males with the best territory, namely areas with lush emergent vegetation.

Female northern green frogs lay 1,000 to 5,000 eggs, which hatch in three to five days. Tadpoles that emerge in the spring metamorphose by the end of the first summer while those hatching later overwinter as tadpoles and metamorphose the following year. Tadpoles eat plants; adults consume a variety of invertebrates and smaller vertebrates, namely fish and other frogs. Adult green frogs, as bullfrogs do, overwinter underwater in the muck and debris of the pond or pool's bottom. Green frogs share many of the same predators as bullfrogs, including humans hungry for frog legs. When startled from the shore, green frogs emit an alarm call, an urgent, high-pitched chirp, and dive into the water.

Northern leopard frogs (*Lithobates palustris*) are medium-sized frogs. They are green, brown, or tan with black or brown spots in two or three rows down the back. The round or square spots are suggestive of the patterns on leopards. A noticeable dorsolateral fold runs most of the length of the frog, from the eye to the hind legs. Leopard frogs often wander from ponds into nearby fields, hence many people call them "meadow frogs" or "grass frogs." Their colors blend in with vegetation. Female northern leopard frogs lay between 300 and 6,000 eggs, which hatch in about one to three weeks, depending on water temperature. In about three months, the herbivorous tadpoles metamorphose. Adults eat a variety of invertebrates as well as smaller frogs. A wide variety of reptiles, amphibians, mammals, and birds prey upon northern leopard frogs. Northern leopard frogs make slow, snoring calls.

The wood frog (*Lithobates sylvaticus*) is the earliest to breed in Ohio, seeking vernal pools as early as late January in some parts of the state. Wood frogs migrate en masse to woodland pools that may be mostly covered in ice. During courtship, they emit a series of short, medium-to-low-pitched clucks. They are smallish frogs, ranging from about one-and-one-half inches to just under three inches in length, and they are predominantly red-brown or red-tan colors, although their hues do vary. Their black masks, just below and extending behind their eyes, are quite conspicuous. The mask is often called a "robber's mask." Dorsolateral folds are present but not as noticeable as on, say, green frogs.

Wood frogs are northern frogs, extending into northern Canada and Alaska, and are common throughout Ohio. Unlike other true frogs, wood frog adults spend little time in or near water. They spend most of the year hidden under rocks and logs of the forest. Their coloration allows them to blend in with leaf litter. The diet of wood frogs consists of invertebrates such as spiders, beetles, crickets, worms, slugs, and snails.

In late March, wood frogs are bountiful in vernal pools. During the first warm (a relative term) rainy night of the spring, if you venture to a vernal pool, you are likely to see many—sometimes hundreds—of wood frogs swimming and calling. Sweep a spotlight across the water's surface, and you will find some floating, some swimming, and some mating pairs embraced in amplexus. The females leave 500 to 3,000 eggs in small, globular masses, which the male fertilizes. Large masses of eggs help retain heat and accelerate development. Return two to three weeks later, and you will find these eggs hatching and tadpoles emerging, dark green with gold flecks on their backs and iridescent bellies.

Because of their need for fishless breeding grounds, wood frogs are an important indicator species for vernal pools. Destruction of forested habitats can have dramatic impacts on local species, so, while vernal pools are sometimes saved during construction, the loss of surrounding

Gray treefrog tadpoles are unable to eat until their mouthparts develop. As they mature, the larvae resorb their tails, conserving energy.

woodlands can destroy wood frog populations. And, although secondary growth forests are increasing in Ohio, wood frog populations are slow to return to disturbed habitats.

All six species of treefrogs (family *Hylidae*) found in Ohio may be seen in ponds and vernal pools. Treefrogs tend to have thinner bodies and long hind legs as compared to those in the true frog family. Some, but not all, treefrogs have suction discs at the ends of their toes. Ohio's treefrogs include the gray treefrog, Cope's gray treefrog, Blanchard's cricket frog, the mountain chorus frog, the western chorus frog, and the northern spring peeper.

The common gray treefrog (*Hyla versicolor*) and its nearly identical relative, Cope's gray treefrog (*Hyla chrysoscelis*), are secretive creatures and well camouflaged, making them nearly impossible to spot even when they are vocalizing nearby. The calls are high-pitched trills, slightly faster for *H. chrysoscelis* (although the speed of each species varies with the temperature). Besides their calls and internal differences, the two frogs look and behave similarly, so they will be discussed together here.

Gray treefrogs are about one to two inches long and variable in color, changing to blend in with their surroundings. They range from gray to green, with dark patterns outlined in black on their backs. Since their skin is somewhat dry and bumpy, they were once called tree toads. The bellies of gray treefrogs are dirty white, and the insides of their hind legs are bright yellow.

Gray treefrogs are arboreal for most of the year. Each spring they migrate to ponds and vernal pools for mating. Females lay up to 1,500 eggs in small groups of thirty to forty. These hatch in four to five days, and metamorphosis occurs in six to nine weeks. Juveniles feed on plants, and adults feed on small invertebrates. Snakes, birds, and small mammals attempt to prey on gray treefrogs. When discovered, gray treefrogs leap, displaying the bright yellow on their legs. The flashy color is hidden again upon landing, perhaps confusing predators. *H. versicolor* ranges across the entire state while *H. chrysoscelis* is found only in the southern third.

Blanchard's cricket frog (*Acris crepitans blanchardi*) is a small treefrog that clearly does not live up to its family name, for it never inhabits trees. Typically it is found along the shorelines of ponds and other permanent bodies of water. It might be found at a vernal pool if permanent waters are nearby. Blanchard's cricket frog is only one-half inch to one-and-one-half inches long, and it varies widely in color, from brown to tan to olive. Like the gray treefrogs, Blanchard's cricket frogs can change colors. And like other treefrogs, they have small sticky discs on their feet.

Perhaps the easiest way to identify the Blanchard's cricket frog is from its call, which sounds like tapping two marbles together, making a clear, loud clicking noise. Calling late into the summer sometimes, the males attract females to their closely defended territories. Females can lay up to 250 eggs, which are attached singly or in small clusters to submerged vegetation and hatch within a few days. Some amphibians are actually larger before metamorphosis than as mature adults, and Blanchard's cricket frogs are among these. Most adults die after breeding, which occurs the year after hatching. Adults eat mostly insects. Their range covers the western half of the state.

The mountain chorus frog (*Pseudacris brachyphona*), one of three chorus frogs within the family of treefrogs, are larger than Blanchard's cricket frog, ranging in size from one to one-and-one-half inches long. They have smooth skin that varies from light brown to olive. Their bellies are off-white. A wide black stripe runs from the eye toward the back. These woodland treefrogs are found in the far southeastern part of the state, where they seek vernal pools and other fishless waters each spring. The male's call is an ascending "raking" sound, like running your fingers across the teeth of a comb. Other than during breeding season, mountain chorus frogs are seldom seen. Females lay 300 to 1,200 eggs in small clusters and then leave the next day. The eggs hatch in three or four days, and metamorphosis occurs in about two months. Habitat disturbances, especially surface mining within its range, may have eliminated some populations.

Western chorus frogs (*Pseudacris triseriata*) are small treefrogs, ranging in size from less than an inch to one-and-one-half inches. Similar in size and breeding habitat to spring peepers, they can be distinguished not so much by their coloration of brown, gray, olive, or tan but by the three stripes, usually solid but sometimes broken into spots, that run down their backs. Their calls are easily confused with the mountain chorus frog because they both produce a raking sound, except that the western chorus frog's vocalization is a bit lower and slower. They migrate early, even when snow and ice remain, breeding in vernal pools and other shallow waters. Females lay up to 1,500 eggs in masses of about a couple dozen to around 300, and these hatch within a week. Metamorphosis occurs in two to three months. Western chorus frogs, originally a prairie species, are found throughout most of the state due to the demise of our forests. After breeding season, western chorus frogs hide in cool, moist places and are rarely seen.

Northern spring peepers (*Pseudacris c. crucifer*) are true harbingers

Western chorus frogs breed in shallow vernal pools. A prairie species, they migrated east with the cutting of Ohio's forests.

of spring. These coin-sized chorus frogs, sometimes a bit over an inch long, are colored in various combinations of yellow, brown, tan, and olive with dirty-white bellies. They can be most easily identified by the large black "X" on their backs (*crucifer* means "cross-bearing"). Sticky toe pads are large and quite noticeable.

Males establish territories about a foot in diameter and begin calling early in the spring, often before ice has completely melted. Congregating in shallow waters such as vernal pools, males let out their surprisingly loud whistles, which are short and up-slurred. Their calls can be heard at great distances, and, if many individuals are calling at the same time, their sound can be deafening. Up to 1,000 eggs are individually attached to submerged plants. The larvae hatch in three to fifteen days, metamorphosing in June. Spring peepers are distributed across the state, although suitable habitats are less common in western Ohio.

Of the three toads found in Ohio, two of them come from the family of true toads (family *Bufonidae*). True toads have prominent parotid glands right behind the eyes, and these secrete a toxic chemical called bufotenins. Anyone who has seen a dog lick a toad knows how effective these toxins can be on predators. Dogs begin producing large volumes of frothy saliva after contact with the bufotenins. True toads have dry, warty skin, which, despite folklore, does not cause warts on humans. These toads are terrestrial, returning to ponds and vernal pools only for breeding.

The first of the two true toads is the American toad (*Anaxyrus a. americanus*), which ranges from two to three-and-one-half inches long. They tend to be tan or brown, although other colors often occur, including olive, gray, black, and red. The underside is cream-colored with spots on the chest. Black spots cover their backs, with one or two warts within each spot. Females are larger and more starkly patterned than males.

In the spring, the male's high-pitched trills can be heard from a variety of wetlands. Calls can last as long as twenty seconds each. Females lay long strings containing up to 20,000 gelatinous eggs, which hatch after three or four days. Black tadpoles school in shallow water, metamorphosing in June, when large numbers of tiny toads can be seen hopping near breeding sites.

To survive winter temperatures, these terrestrial amphibians burrow deep enough to avoid freezing. While the larvae are herbivorous, adult toads are welcome additions to gardens, where they consume insects, both larvae and adults, as well as a variety of other small invertebrates. While some animals are deterred by the deleterious effects of the parotid toxins, certain species of snakes and birds still consume them. In response to threats, toads emit a chirp, which presumably startles predators, and

A gray treefrog peers out of a pitcher plant. Pitcher plants are found in bogs and fens.

they puff themselves up, perhaps appearing too large to be swallowed. They also urinate upon being disturbed, so be forewarned—if you pick one up, you may find a bladderful in your hands!

The second true toad is the Fowler's toad (*Anaxyrus fowleri*), which is quite similar in appearance to the American toad. Fowler's toads tend to be found within river and stream valleys and in sandy uplands. Few are found in the northeastern portion of the state, except in a narrow band all along Lake Erie.

Generally, Fowler's toads can be differentiated from American toads in several ways. First, the bellies of Fowler's toads do not have spots. Second, whereas the American toad has large warts on the back legs,

Fowler's do not. Third, upon close inspection of the spots on their backs, Fowler's toads tend to have three or more warts within each black patch while American toads tend to have one or two. Finally, and most notably, their calls are different. While American toads have an enduring trill, Fowler's toads let out a slightly nasal, whirring "w-a-a-a-h," which lasts three to four seconds.

While Fowler's toads are also terrestrial, they migrate to all types of waters for breeding throughout spring and into summer. Like the American toad, Fowler's toads lay eggs in strings, in their case numbering up to 8,000. Larvae hatch in two or three days and metamorphose in about a month. Adults eat a diet similar to adult American toads, and they, too, hibernate in underground burrows. Both are found commonly across the state, even in urban areas if breeding sites are nearby. Fowler's toads defend themselves with parotid glands, but they also can flee from predators. Unlike the plodding hops of American toads, Fowler's toads speed to safety using rapid, scooting hops. American and Fowler's toads can interbreed.

The eastern spadefoot toad (*Scaphiopus holbrookii*), about two inches long, has relatively smooth, brown skin with some red warts. Its belly is white or gray. The call of males is a harsh, crescendoing "whar," which can be heard when the eastern spadefoots emerge from their underground burrows for breeding.

These toads are the only Ohio amphibians without a definite breeding season, tending to emerge after heavy rainfalls during the warmer months. They congregate in temporary pools, where females lay bands of eggs, one to two inches wide and about a foot long, which hatch in two or three days. Metamorphosis can occur in as little as two weeks or as long as nine weeks. Sometimes breeding occurs multiple times within the same year, and other years it may not take place at all. After breeding, eastern spadefoot toads are rarely encountered. A small population of eastern spadefoot toads occurs in the southeastern portion of the state. They are a state-listed endangered species.

Salamanders

Nine types of salamanders can be found in and around Ohio's ponds and vernal pools. Salamanders differ from frogs and toads in that they have tails as adults, do not call (perhaps an occasional, quiet hiss is the loudest they get), and appear elongate. Those most commonly found in shallow ponds and vernal pools are the mole salamanders (family

Ambystomatidae). All *Ambystoma* species have stout bodies, breathe by means of well-developed lungs, and spend the majority of their lives underground—in self-made burrows, in excavated tunnels of other animals, or wedged under logs, stones, and other suitable cover. Many of them are so locally common that during breeding season hundreds may be seen in one night, yet you might not see a single one any other time of the year. For Ohio, five of the six indicator species of salamanders are from this family.

The best-known mole salamander and symbol of vernal pools is the spotted salamander (*Ambystoma maculatum*). These flashy salamanders have bright yellow spots on their bluish-black or brownish-gray backs. These spots are arranged in two rows, which typically converge as one on their tails. Their bellies are some shade of gray. Spotted salamanders grow to about eight inches, with typical adults measuring between four and six inches.

Spotted salamanders spend much of their lives underground, either under objects such as logs and stones or in borrowed or self-made burrows. But each spring, spotted salamanders venture forth during the first warm rains of March to return to ancestral pools. Most return to the same one in which they were born. These salamanders may travel hundreds of feet to vernal pools and shallow ponds, where they congregate en masse—in what is called a "congress"—about the same time wood frogs gather together. Sometimes hundreds of salamanders dance within the pools, wriggling up from the bottom, breaking the water's surface, and diving back to the bottom again. This dance is used to attract mates, but it is also useful for individuals to obtain sufficient oxygen, especially due to low levels of dissolved oxygen in the cold water and the high level of energy required for courtship, mating, and egg laying. Males deposit individual spermatophores on the bottom of the vernal pool; then females draw up seminal fluid from each and fertilize up to 200 eggs. They lay their eggs in masses of up to 100. These hatch in as little as three weeks.

Larvae have external gills, which are resorbed during metamorphosis, typically in about three months, although larvae can change more quickly or more slowly depending on their environmental conditions. For example, if larvae sense that a vernal pool is drying quickly, they accelerate metamorphosis. Larval size varies depending on the depths of their pools and their supplies of food. Spotted salamanders are found across the state in wooded uplands with breeding pools nearby. Flat topography and habitat destruction for agriculture in northwest Ohio are the probable cause for far fewer populations there. Spotted salamanders are a vernal pool indicator species in Ohio.

The abundant Jefferson salamander (*Ambystoma jeffersonianum*) is similar to the spotted salamander in shape and size, except it has silver-blue flecks peppering its sides. Jefferson salamanders migrate early, appearing in late February in some locations. Secretive and *fossorial* (i.e., they burrow), Jeffersons can be seen in great numbers during their spring migration, where they, too, congregate and dance in vernal pools and fishless ponds. Females lay about 200 eggs in small masses. These hatch in about a month and metamorphose in two or three months, something, again, quite dependent upon environmental conditions. Jefferson salamanders are common south and east of a line running approximately from Cleveland through Dayton. They, too, are vernal pool indicator species.

The abundant small-mouthed salamander (*Ambystoma texanum*) ranges in size from four to seven inches in length and is named for its relatively small mouth. Its tail, back, and small head are dark gray, brown, or black and are speckled with grayish splotches. Their bellies are dark as well. These salamanders breed in a variety of waters, from quarry pools and vernal pools to ditches and flooded lawns. Breeding can occur in winter, sometimes as early as December in southern Ohio. Pool-breeding females lay between 300 and 700 eggs, and larvae emerge in one to three

A marbled salamander larva is ready to metamorphose into a young adult. Mainly breathing air now, it will leave the pond within days.

months, most often in April. They metamorphose in seven to nine weeks. Small-mouthed salamanders are found in much of the state except the unglaciated Allegheny Plateau to the east and south.

Marbled salamanders (*Ambystoma opacum*) are striking in their appearance, black with white or gray bands on their backs. They grow to almost five inches in length and are distinct among Ohio's mole salamanders for breeding in fall. Although small in size, their early breeding gains them a competitive edge. Females seek vernal pools that are nearly dry but will hold water until midsummer. Females lay about 100 eggs under the pool's leaf litter and then wait with them until fall rains cover them. The larvae emerge in the fall anywhere between two and nine weeks after egg-laying. Metamorphosis occurs near the beginning of summer, again depending on environmental conditions. Marbled salamanders tend to inhabit the southern half of the state and are indicator species of vernal pools.

The eastern tiger salamander (*Ambystoma tigrinum*) is the largest salamander inhabiting Ohio's ponds and vernal pools. It grows to almost ten inches in length and ranges in color from blackish-brown to almost black with large yellow to tan randomly placed blotches on the head, back, and tail. Their bellies are light gray. Tigers are able to dig their own burrows, staying underground most of the year. They emerge in massive migrations to large vernal pools in the spring, where they participate in energetic courtship rituals for only a few nights. Females lay several hundred eggs in a handful of masses. The larvae emerge in one to two months and metamorphose in three months. The larvae are voracious predators. Generally, tiger salamanders are found north and west of a line running from Cleveland through Cincinnati. Eastern tiger salamanders are considered an indicator species for Ohio vernal pools.

Finally, residents of the extreme northwest counties might come across a rare mole salamander, the blue spotted salamander (*Ambystoma laterale*), which looks something like the Jefferson salamander except with many more blue flecks. It looks like an old-fashioned piece of enamelware. Blue spotted salamanders breed in vernal pools around April, with the female laying up to 500 eggs, which she attaches singly or in masses up to fifteen on aquatic vegetation.

In addition to the previously mentioned *Ambystoma* species, past hybridization has resulted in many populations of unisexual individuals. These look similar to their ancestral parents but are impossible to assign a species name.

Another family of Caudata found in Ohio are the lungless salamanders (family *Plethodontidae*), of which only one species is found in

Tiger salamander larvae are voracious eaters, consuming large numbers of amphibian larvae, including their own kind.

Ohio's ponds and vernal pools. Lungless salamander do not have lungs (and they do not have gills, either). Instead, they breathe entirely by absorbing oxygen through the skin or the lining of the mouth.

The four-toed salamander (*Hemidactylium suctatum*), growing to only three-and-one-half inches in length, is the smallest salamander found in Ohio. The four-toed salamander is so named for its unusual back feet, which have four instead of five toes. (Most salamanders have four toes on their front feet, which is also the case for the four-toed salamander.) Its belly is strikingly white with scattered peppercorns, and it is typically ornamented with rusty-red coloration above and grayish coloration on the sides. The base of the tail is constricted. If carelessly handled, they lose their tails.

Four-toed salamanders require undisturbed woodlands with mossy bogs. They mate in the autumn or early winter, prior to migrating to breeding sites. In the early spring females migrate to nesting sites, where they lay about forty eggs in individual or collective nests, often depressions in mosses. Females remain with the eggs during incubation (usually only one or two females stay with each collective nest). Larvae

A red eft form of the red-spotted newt wanders through the forest. Within a few years it will return to water and transform into an aquatic adult.

emerge in one or two months, and metamorphosis occurs four to six weeks later. In Ohio, the four-toed salamander is fairly widespread but rare, hence its threatened species status, and it is considered an indicator species for vernal pools.

Only one newt (family *Salamandridae*), the three- to four-inch-long red-spotted newt (*Notophthalumus v. viridescens*) is found in Ohio. Newts have skin that is warty rather than smooth, and they produce skin toxins. Whereas most salamanders have three distinct stages—egg, larva, and adult—red-spotted newts have four. These are egg, aquatic larva, terrestrial juvenile (the red eft stage), and adult. Terrestrial red efts are red-orange with bright red spots bordered in black. Adult red-spotted newts are olive-colored with the same red and black spots. The bellies of both stages are light yellowish. Larvae emerge from their eggs in ponds and vernal pools but soon lose their gills and move to land, where they live for two years. They then metamorphose into the adult stage, returning to live their lives out in the water. Females lay 100 or more eggs, which hatch in three to five weeks. Red-spotted newts are common in Ohio, especially in the northeastern and southern parts of the state.

Turtles

Nine types of turtles swim in and crawl around Ohio's ponds and vernal pools. The common snapping turtle (*Chelydra s. serpentina*), the largest of Ohio's turtles, can grow to over fourteen inches long and in excess of thirty-five pounds. The snapping turtle's carapace is tan, green, olive, or black and has knobby or pointed *scutes*. The head and tail are greatly enlarged, and while stories of the snapping turtle's bite are sometimes exaggerated, great care should be taken in handling them. Larger specimens are considered quite dangerous and should be left alone. Females lay about two to four dozen eggs. Interestingly, if an egg incubates at about seventy-seven degrees Fahrenheit, then a male will result. Temperatures above or below will result in females. Temperature may vary from the top to the bottom of the nest, producing both sexes. While snapping turtles are common throughout the state, the fact that snapping turtles do not bask much results in few spottings.

The most common turtle seen basking on logs is the midland painted turtle (*Chrysemys picta marginata*). This decorative turtle has a brown-green *carapace,* up to ten inches in length, with red and black patterned marginal plates. They consume a variety of aquatic plants and animals, such as insects, leeches, aquatic earthworms, snails, fish, and frogs, the last two probably as carrion. Females produce between four and fifteen eggs per year. Cooler turtle eggs produce males and warmer eggs produce females. In late autumn, midland painted turtles bury themselves in the muddy bottom and obtain oxygen through absorption along the inner linings of their mouths and cloacae.

Bad-tempered snapping turtles are equipped with sharp jaws. Contrary to popular belief, their diet consists of carrion and plant material.

The threatened spotted turtle (*Clemmys guttata*) is a three- to five-inch turtle with a black carapace dotted with small, bright yellow spots. It is sometimes found along pond edges, but it is most often seen in peat lands, where it forages under the bog mat. Females lay one to eight eggs from May to July.

The four- to five-inch-long common musk turtle (*Sternotherus odoratus*) has a dark brown domed shell. The brightest coloration is the pair of yellow stripes on either side of the head. Also known as stinkpots, these turtles release a foul odor when disturbed. Females typically lay two to five eggs near water. Found across the state, stinkpots prefer deeper ponds. Believe it or not, they are also arboreal, sometimes falling on the heads of unsuspecting canoeists!

Northern water snakes prey on frogs, crayfish, and other small pond animals. Young individuals are boldly marked with broad bands.

Snakes

Three species of Ohio snakes are commonly found near ponds and vernal pools. The common water snake (*Nerodia s. sipedon*), which can grow to three-and-one-half feet in length, has patterns of black, brown, and red bands across its body. Northern water snakes quickly retreat when approached; when captured, however, they are notoriously aggressive, biting almost immediately. Northern water snakes are top pond predators, consuming various invertebrates, amphibians, fish, birds, and mammals.

Eastern garter snakes (*Thamnophis s. sirtalis*), sometimes mistakenly called "garden" snakes, can exceed in two feet in length and are generally brown with three yellow stripes. Stunning patterns of green, blue, brown, and orange are common. While strongly terrestrial, garter snakes can often be seen basking near ponds or venturing into the waters along the edges to feed on amphibians and other pond creatures.

Related to the garter snakes are eastern ribbon snakes (*Thamnophis s. sauritus*), which grow to about the same size as their *colubrid* relatives, but ribbon snakes have exceptionally long tails, which range from one-quarter to one-third their entire body lengths. Ribbon snakes are semiaquatic, eating toads, frogs, salamanders, tadpoles, and fish. They seldom venture far from bodies of water. Like garter snakes, ribbon snakes are excellent climbers. Ribbon snakes are found across the state except in the southwest.

What Fur?

Mammals

O f the mammals seen around Ohio's ponds and vernal pools, some can be considered permanent residents, and others are frequent visitors, occasionally visiting the waters to eat and drink. Five are residents, the first of which builds its own ponds: the American beaver *(Castor canadensis)*. Beavers are unique animals in that they can control their own environments. After finding suitable locations—such as slow-moving streams—beavers, usually in small, familial colonies, begin constructing dams with sticks and mud. Typical dams are three or four feet high and 50 to 200 feet long. Impounded waters are referred to as beaver ponds. Within a beaver pond, beavers construct a lodge with two or more underwater entrances.

Beavers are Ohio's largest rodents, weighing up to seventy pounds. Their dense and beautiful fur is dark chestnut brown. The beaver's wide, flat tail is covered with scales, an indication of its reptile ancestors. Along with the tail, webbed hind feet help make the beaver an efficient swimmer. Underwater, valves on the ears and nose close, yet beavers swim with their smallish eyes open. Besides the flat tail, the most noticeable feature of a beaver is the prominent set of incisors protruding from its mouth. These are used to chew trees for food and for dam and lodge construction. Large lungs allow beavers to stay underwater for as much as fifteen minutes. Beavers are monogamous, mating for life. Small litters of about four are born each spring, and young stay with parents for two years.

Beavers were once common throughout Ohio. When Europeans settlers arrived, massive fur trading began. By 1830, beavers had been extirpated from the state. Today, thanks to careful management practices, beavers have returned to the state and are common in over half the state, especially the eastern and southeastern counties. The best time

Opposite: The closing of autumn marks the end of the pond year. With the onset of cold weather, amphibians, reptiles, and other animals begin a period of hibernation. Many smaller organisms simply perish.

Beavers are well adapted for aquatic life. They have valves that close their ears and nostrils along with dense fur, which protects them from freezing pond water.

to see beavers is just before sunset. Sit quietly along the water's edge and wait for activity. But don't be surprised if, after a beaver spots you, it loudly slaps its tail on the water's surface, the beaver alarm signal, which indicates potential threats.

Another lodge-building rodent widespread in Ohio ponds is the muskrat (*Ondatra zibethicus*), which has dark brown fur and a long, ratlike, scale-covered tail. Much smaller than beavers, muskrats grow to two feet long, but half of that is tail. Their semiwebbed hind feet and, especially, their tails aid in swimming and allow muskrats to reach speeds of three miles per hour. In some locations, typically marshes, muskrats build large, domed lodges, which can be mistaken for beaver lodges. In pond environments, muskrats often dig burrows in the shores surrounding the water. These tunnels can lead to the significant erosion problems.

Muskrats live in families of one male, one female, and their young, and they defend their territories aggressively. Two to three litters of six to eight young are born each year. These rodents feed primarily on aquatic vegetation, but they do consume aquatic invertebrates, fish, frogs, and turtles. Muskrats, like beavers, have been trapped extensively for their fur. They are widespread across Ohio. While muskrats are best seen at dawn and dusk, some may be seen during the day.

Another unique inhabitant of ponds is the star-nosed mole (*Condylura cristata*), which reaches six to eight inches in length and has blackish-brown, water-repellent fur and a long, stout tail. Its nose is adorned with twenty-two mobile, fingerlike tentacles covered with almost 100,000 touch receptors known as *Eimer's organs*, which help in identifying prey for this blind insectivore. Star-nosed moles are semiaquatic, burrowing along the edges of ponds. Their tunnels often extend and exit underwater. Besides insects, the star-nosed mole eats other invertebrates, including worms and crayfish as well as small fish. Their ability to smell underwater—by breathing out air and then drawing it back in—also aids in feeding. They breed once a year, typically producing four or five young. Star-nosed moles are one of the few warm-blooded animals that may be active during the winter. They are found in the northern half of the state on the edges of ponds, lakes, rivers, and streams and in marshes and wet meadows.

If you are lucky enough to spot river otters (*Lontra canadensis*), you will likely enjoy watching them play, twisting in the water and wiggling

An abundance of muskrats can critically impact a pond's ecosystem. Overpopulation results in "eat-outs," destroying much vegetation, thus impacting other inhabitants.

up the shore. River otters grow to around three-and-one-half feet long, about one-third to one-half of which is tail. Otters are covered in a dark brown fur that is sleek and glistens when wet. Their long, streamlined bodies, short fur, completely webbed feet, and powerful tails make them exceptional swimmers. The long whiskers of the river otter aid in detecting and capturing prey, which includes fish, frogs, crustaceans, and sometimes mammals. Some may occasionally attack large birds such as ducks, geese, and herons. The name of this carnivore is misleading, because it lives in a wide variety of environments, ranging from marine to freshwater and from lakes to streams. River otters prefer larger bodies of water, so it is uncommon to see them in ponds.

Mink (*Mustela vison*) may be found in good-sized ponds. They grow to about two feet in length, not including a bushy tail about one-third the length of the body. They are covered in a rich brown fur with a white patch under the chin. Semiaquatic, minks build their dens along the banks and shores of rivers and lakes. They are excellent swimmers, diving to prey on fish, crayfish, frogs, and turtles. They also feed on land, eating eggs, small mammals, snakes, ducklings, and other vertebrates. Quite territorial, male minks hiss, snarl, and screech while aggressively—and sometimes mortally—attacking intruding animals. They breed in the late winter and give birth to two to ten young in the spring. Mink fur is highly prized, leading to trapping and decreased populations. While mink were probably common in Ohio, trapping has reduced

The largest of our native rodents, beavers can weigh up to seventy pounds.

their numbers. Today mink are found in all eighty-eight counties, with greater numbers in the eastern and southeastern regions.

A frequent visitor to ponds and vernal pools is the raccoon (*Procyon lotor*). The infamous masked mammal grows to eighteen to twenty-eight inches long and can weigh up to thirty-five pounds. Their fur is a gray and black with some brown mixed in. The black mask with white above and below and the rings of yellowish-white and black on the tail make this mammal easy to identify. Raccoons are omnivores, eating anything from fruits, nuts, and grains to insects, frogs, turtle eggs, and crayfish, which they sometimes dunk in the water before eating. Raccoons breed in late winter and have two to seven young in early spring. These curious mammals are widespread and commonly seen in urban as well as rural habitats across the state.

The last of the mammals that frequent Ohio ponds are found not in or even around them; rather, they are found *above* ponds. These are several species of bats. Ancient mammals, having evolved some 50 million years ago, bats are the only true flying mammals, having enlarged hands with double membranes of skin stretched across the finger bones as well as skin stretched between the legs. All bats are nocturnal, and those found in Ohio feed on insects caught while in flight over the water and surrounding shores. Bats have poor eyes, which are not their primary means of sensing. Instead, they emit a series of high-pitched sounds—too high for humans to hear—which bounce off objects and return to the bat's ears. This method of echolocation allows bats to "see" prey and other objects, similar to the way sonar works. Bats suspend themselves upside down by day to rest.

Bats typically mate in the fall, right before hibernation. Ovulation and fertilization are delayed until the end of hibernation. Females give birth to one or two young, which hang onto their mothers until they become too heavy. They then stay at the roost during feeding. Several bats in Ohio are commonly seen swooping over ponds and vernal pools at dusk. The little brown bat (*Myotis lucifugus*) is two-and-one-half to four inches long, has a wingspan of about six inches, and weighs only as much as a nickel. A bit larger is the big brown bat (*Eptesicus fuscus*), which is four to five inches long, has a wingspan about double that of the little brown bat, and is twice the weight. Both brown bats are brown with dark brown wings. The silver-haired bat (*Lasionycteris noctivagans*) is about the same size as the little brown bat and covered with frosted black hair. Rabies among bats has been greatly exaggerated, and bats, therefore, are unreasonably feared. In fact, insect-eating bats are quite beneficial in controlling pests.

Finely Feathered Fauna

Birds

During a midyear visit to a pond, you are likely to see a variety of animals zipping through the air above summer-warmed waters. Many insects buzz in every direction, and closely following them will be predators, birds hungry for a meal. Some avian animals—such as herons or hawks—select other fare, keenly searching for many forms of aquatic prey. Vegetarians may be dabbling along the pond's surface, wetland teeter-totters dipping down for aquatic plants. Whatever is on-wing when you visit, you can enjoy the favorite pastime of birdwatching.

One of the most common birds found around ponds and larger vernal pools are ducks, members of the large waterfowl family *Anatidae*, which includes dabbling ducks and diving ducks as well as geese and swans. Most common of the ducks are dabbling species (subfamily *Anatinae*), which are at home in the shallow waters of ponds. Dabbling ducks are so called because they poke their bills into the water or turn their heads downward from the surface to pick at or "dabble" on aquatic vegetation. Some feed terrestrially as well. With their legs centered under their bodies, dabbling ducks can walk well on land. Strong fliers, dabbling ducks can take off from water without running, and most are migratory.

The most common of the dabbling ducks by far is the mallard (*Anas platyrhynchos*). Male mallards are hard to mistake from their nondescript female counterparts. The male's mating plumage includes a bright green head and neck with a white ring just below, a rust colored breast, and a pale body. Females are more nondescript, basically brown with a black eye line and a metallic-blue speculum or wing patch. Females are most easily identified in association with males. Mallards feed largely on seeds. They sometimes nest far from the water's edge. Widespread, mallards can be seen year-round in pairs or small groups on ponds.

Opposite: A top predator in the food chain, great blue herons are common visitors to Ohio wetlands.

109

Most mallard ducklings will not survive their first month. Snapping turtles lurk below the water's surface. Northern water snakes snap up juveniles. Some are even attacked by giant water bugs.

Because wood ducks (*Aix sponsa*) prefer sheltered waters among trees, they are often found in secluded ponds and larger vernal pools. They are usually found in pairs or small groups. Breeding males of this dabbling duck are one of the most beautiful birds in the state, with green, red, black, and white on their heads, brown, blue, yellow, and white on their bodies, drooping crests, and red eyes. As with most birds, females have less distinctive, duller brown-gray colors but with a noticeable white patch around the eyes. The diminutive size of wood ducks, about eighteen inches long, and their tendency to inhabit quiet waters helps in identification. Wood ducks reside in all eighty-eight counties during the summer and migrate south for the winter. They are considered uncommon. Wood ducks feed on acorns and other seeds and nest in cavities in trees. People often attract wood ducks by placing nesting boxes in appropriate habitats.

One small diving duck commonly seen on farm ponds and vernal pools is the hooded merganser (*Lophodytes cucullatus*). Males in breeding plumage are easily recognized by the prominent white patch in the middle of their black crests, their blackish bodies, and reddish-brown wings. Females are brown and gray with a frosted brown crest.

This dabbling duck feeds on fish as well as invertebrates such as insects and crayfish. When flying, the wings of the hooded merganser produce a high-pitched trill. Like wood ducks, hooded mergansers nest in tree cavities. While sometimes year-round residents in Ohio, they are more often seen passing through during spring and fall migrations.

The Canada goose (*Branta canadensis*) is one of the most widely seen waterfowl across the state. With its black head and neck, gray body, white breast, black feet, and a wingspan of about five feet, it also is unmistakable. In early spring, geese build nests several feet across just above the water. Canada geese glean leftover crops from fields, roam on golf courses, and stroll through wide-open lawns, usually near water. While geese are enjoyable to observe as they honk and fly gracefully in large flocks, landowners often consider them a nuisance due to the extent of their grazing, the unpleasant waste they leave behind, and their aggressive behavior.

A few grebes (family *Podicipedidae*) can be found in Ohio. Grebes are excellent swimmers and divers, with feet set far back on their bodies for strong propulsion during swimming. Besides when they are nesting, grebes are rarely seen on land. The pied-billed grebe (*Podilymbus podiceps*) is the most likely species to be seen on ponds. This grebe is quite small and generally light brown in color. Distinctive markings on

Canada geese bring their families to ponds in the spring.

breeding males include black rings around their otherwise white bills, white eye rings, black caps and throats, and a white patch at the rump. They feed on insects, crayfish, and fish. Uncommon but found across the state, pied-billed grebes tend to be solitary or in small flocks.

While several members of the rail (*Rallidae*) family can be seen across Ohio, the most commonly spotted on ponds is the American coot (*Fulica americana*). Adults are generally dark gray with black heads and a swollen red spot or frontal shield above the base of their white bills. These smallish, ducklike birds dive for aquatic vegetation, which they eat while bobbing back and forth on the surface. Coots often congregate

Hard to detect, a sora skulks through shoreline vegetation.

on ponds and in marshes in large, tight flocks, and their calls, chicken-like clucks, can create quite a racket. While few nest in Ohio, American coots are quite common during migration.

Another rail, the sora (*Porzana carolina*), often inhabits marshes but sometimes visits ponds. They may be seen moving among the shoreline vegetation. Soras are eight to ten inches in length and gray-brown in color. They have short yellow bills and black spots on the face and throat. Soras are omnivores, with diets that include, among other things, seeds and various invertebrates. More often heard than seen, the sora makes a variety of sounds, including a descending "wepwepwepwepppppprrr" and, if alarmed, squealing and clattering sounds.

Among the various wading birds, herons (family *Ardeidae*) are the most common around ponds and even bigger vernal pools. These relatively large birds have coiled necks and pointed beaks, both instrumental in striking prey. The most common of these birds and the symbol of Ohio wetlands is the great blue heron (*Ardea herodias*). More gray than blue, this heron is quite large, over four feet in length and with a wingspan of six feet. The long, trailing crest on the head is distinctive. When searching for prey along shores, great blue herons stand tall and often stationary, quickly spearing tadpoles, frogs, fish, and even small mammals with their strong bills. Great blue herons nest in rookeries, where dozens of pairs congregate each spring—an incredible sight to behold. Their calls are loud and piercing squawks, sometimes sounding doglike. Great blue herons are common and widespread, residing in Ohio year-round.

The less common green heron (*Butorides striata*) is smaller and often skulks along the shores of ponds. Green herons, a little less than half the size of great blue herons, are slate gray-green with a striped, chestnut-colored neck. When foraging, these herons crouch, sometimes dropping feathers or other objects onto the surface of the water to attract fish. Green herons are solitary nesters. Frequently, green herons are heard but not seen, emitting a sharp "skeew" sound and, when anxious, repeating a low knocking sound. Green herons are summer residents across the state.

A number of shorebirds frequent Ohio's ponds. Among the state's species of shorebirds, two sandpipers and the killdeer (all three from the family *Scolopacidae*) are quite common around ponds. Like most shorebirds, the spotted sandpiper (*Actitis macularius*) has a long bill, but, unlike most shorebirds, it has short legs. The spotted sandpiper is dark above and white below and measures only about seven inches in length. Spotted sandpipers teeter back and forth as they forage along shorelines, poking around for small aquatic prey. Uncommon but widespread, they nest across most of the state.

Killdeer (*Charadrius vociferus*), like the solitary sandpiper, has the typical form of shorebirds—long legs and an elongate bill—although it is a bit larger than most other shorebirds, reaching about ten-and-one-half inches in length. Generally, killdeer are brown above and white below, with distinctive white patches on the head along with white and then black neck rings. While killdeer are shorebirds and may be seen around water, they actually prefer dry, grassy areas much of the time, both for feeding and for nesting. Killdeer build camouflaged nests in open, rocky areas, sometimes even in driveways. The species name, *vociferus*, refers to their loud calls, which they emit when they or their nests are threatened. Most killdeer spend the warmer months in Ohio, with a few remaining throughout mild winters.

Among North American kingfishers (family *Alcedinidae*), only the belted kingfisher (*Ceryle alcyon*) is seen in Ohio, sometimes around ponds. Kingfishers hunt by perching or hovering above the water looking for fish and then diving headfirst to catch them. The belted kingfisher measures thirteen inches in length. They are slate gray-blue above, with a shaggy crest, and have a white collar. Females have rufous on the breast. They nest in cavities in dirt banks. Kingfishers are uncommon but widespread across the state throughout the year.

A common warbler species found around ponds is the common yellowthroat (*Geothlypis trichas*). Adult males have a distinctive black mask. Common yellowthroats crouch with their tails upturned, often in the brushy vegetation along the water's edge. The unmistakable "witchety witchety witchety" call of the common yellowthroat indicates its presence even when dense vegetation hides its bright colors. It is one of a few warbler species that raises two broods during a nesting season.

Among Ohio raptors—predatory and carnivorous birds with sharp talons and hooked beaks—the most common hawk to frequent ponds and vernal pools is the red-shouldered hawk (*Buteo lineatus*), a compact hawk seventeen inches long and with a wingspan of forty inches. As its name suggests, the red-shouldered hawk has an orangish patch on its shoulders and similarly colored bars on the breast. Red-shouldered hawks are found near water, often hunting for reptiles and amphibians as well as small mammals and birds. This *buteo* is uncommon but widespread across the state, a year-round resident in all but the northern counties.

Swallows (family *Hirundinidae*) are commonly seen hunting for insects above ponds. Swallows are short-legged and short-beaked songbirds with relatively long pointed wings. The first of these is the barn swallow (*Hirundo rustica*), a small bird just under seven inches long. It is dark blue-black above and lighter underneath and has a rufous

A tree swallow returns to its nest with a dragonfly. Large numbers of dragonflies are taken by birds.

throat. The barn swallow has a characteristic long, forked tail. It is a graceful flyer. Barn swallows typically build their muddy nests on man-made structures, such as bridges, under the eaves of houses, and on the wooden beams of barns. They forage over ponds, hunting insects in acrobatic flight as well as skimming water along the surface. Barn swallows are common summer residents.

Tree swallows (*Tachycineta bicolor*) are about an inch smaller than barn swallows and have only a slightly notched tail. The upper side of these common birds is a metallic green-blue-black, while the underside is white. Besides catching insects, this aerial songbird also eats berries. Tree swallows nest in birdhouses or tree cavities

The purple martin (*Progne subis*) is the largest of Ohio's swallows, reaching eight inches in length. Dark gray above and lighter below, the purple

In March, male red-winged blackbirds begin staking out territories, displaying their red shoulder patches.

Opposite: The first clutch of redwinged blackbird eggs will hatch in June.

martin has a colored patch on its back that is actually more blue than purple. Once tree dwellers, these swallows now nest almost exclusively in man-made houses. They often forage higher than other swallows.

Among Ohio's sparrows, the song sparrow (*Melospiza melodia*) is the most likely to be found around water. About six inches long and streaked with the brown and reddish-brown typical of sparrows, song sparrows have a dark distinctive dark spot on their breasts where the streaks converge. Small and somewhat unremarkable, they are the most common of the sparrows. Song sparrows are year-round residents across the state. These abundant birds sometimes raise four broods in a nesting season.

If you visit any farm pond in the height of summer, you are sure to see and hear red-winged blackbirds (*Agelaius phoeniceus*). These large songbirds are almost nine inches long and have a wingspan just over one foot. The plain females are covered with rufous-brown streaks, but males are distinctively black with single red and yellow patches on their wings. Among a number of vocalizations, their "kon-ka-reeee" call is easily identified. Red-winged blackbirds build nests in dense vegetation, such as among cattails. These *icterids* are common in Ohio and are year-round residents in some southern counties.

Exploration, Succession, and Pollution

What We Can Do

Fortunately, the Buckeye State has plenty of ponds and numerous vernal pools where we can experience nature. Reading about aquatic habitats and their inhabitants is a fine step toward understanding the natural world, but nothing replaces being there—experiencing the vitality in and around these watery realms. Hopefully this book is a catalyst to greater explorations of ponds and vernal pools near you. If you have no special place to study, look for a close-by habitat. Some people build their own ponds or vernal pools, but for most people, it is enough to find a wetland just down the road. Look for inviting waters nearby. Investigate local parks, remembering to ask officials about park regulations. Or seek permission from a neighbor to dip net, fish, or simply sit back with a pair of binoculars at pond or pool's edge. A friendly smile, an intelligent introduction, and assurances of solid safety practices will likely bring not only permission but also others to join you.

The right equipment will make your investigations more enjoyable and productive. Hip or chest waders allow you to step into the water without getting wet and dirty. Be sure to bring along bowls or a small plastic aquarium to temporarily hold specimens. If you have a microscope, make sure to place a few small drops of water on a slide to examine small organisms. Studying ponds and vernal pools with a partner makes learning easier and more fun as well as offering greater safety when working in and around water.

Begin your studies in the daytime. Arrive early during warm months to find dew-covered insects perched on vegetation. You might be lucky enough to find a dragonfly emerging from its nymph stage. Because invertebrates are cold-blooded, they will be slowed down if not immobile very early on cool mornings, an excellent time for observation. During midday, animals are quite active. Frogs call, and dragonflies and birds

Opposite: Aggressive swamp loosestrife (foreground) is one of our most invasive plants. Invading from the shore, its roots trap sediment and hasten the filling-in of ponds

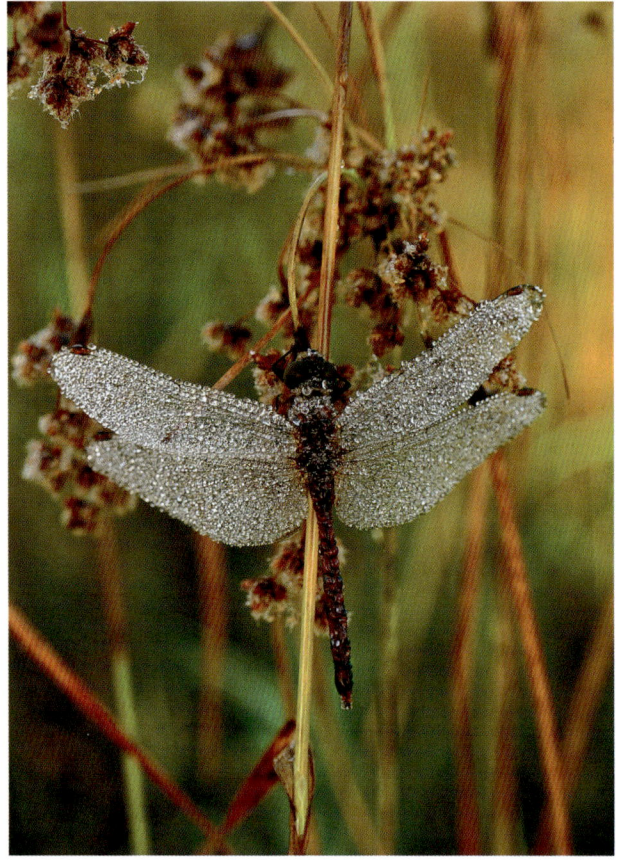

Left: With an air bubble to absorb oxygen, a backswimmer can remain submerged for many hours.

Right: A dew-covered ruby meadowhawk rests for the night. Meadowhawk species are common late-summer dragonflies.

circle overhead. Then treat yourself to the pleasures of advancing night. Just before dusk, sit down in a chair beside the water and listen to the quieting world, which will become increasingly active again with insects calling and silently swooping bats.

Make sure to venture out to your pond or vernal pool in all seasons. Early springtime is great for amphibian activity in vernal pools, and summer ponds are abuzz with dragonflies and damselflies, frogs and turtles, and beetles and birds. In the fall, check out dried bottoms of vernal pools, and in the winter, if the ice is thick enough, walk to the pond's center for a unique view of creatures frozen or slowly moving below. During all times of the year, look around for animal tracks and listen closely for the calls of wild, wonderful wetland inhabitants.

Succession

In studying your pond or vernal pool, watch for changes. Over time, all things undergo transition. Animals grow, and plants flower and produce seeds. Fall brings deciduous leaves to the water's surface, and winter brings many organisms to a near standstill.

Among all these changes, ponds and vernal pools face two major threats to their existence. One is part of the course of natural events; the other is of our own making. First, ponds and vernal pools disappear over time due to succession. This means that they fill in with sediment, eventually becoming more and more shallow. The vegetation around the pond's edges slowly advances toward its center, and robust water plants begin to fill the open water. Consequently, plant and animal populations begin to decline, and aquatic diversity suffers. Some ponds may transform eventually into vernal pools, and some vernal pools into muddy patches.

As more leaf litter falls, more dirt washes in from the watershed, and animal waste collects, even the shallowest pools eventually disappear. This natural process, always aided by the downward pull of gravity, results in succession, which is neither good nor bad from a biological perspective. It is part of the dynamics of nature; from a pond- or vernal pool-owner's perspective, however, succession is not always desired. Fishing ponds become shallower and warmer. Lower levels of oxygen eventually kill off species. Vernal pools no longer fill to a depth sufficient for salamander and frog larvae to develop. In short, animals are forced to move elsewhere. Managers of ponds and vernal pools sometimes choose to reverse succession through deepening basins; others prefer to let nature take its course.

In early summer, midland painted turtles come to the pond's edge to lay their eggs. Most nests are discovered by predators.

Far greater and faster change occurs, however, due to the actions of humans. Habitat destruction and pollution alter all kinds of ecosystems, including ponds and vernal pools. Habitat destruction most often occurs when ponds and pools stand in the way of development. New roads and building construction entail tree clearing, earth moving, and lots of impermeable asphalt and concrete. Trees, vital to detritus-based ecosystems, are removed. Pond and pool basins are bulldozed. And water runs off parking lots and roadways into sewers instead of undergoing natural filtering processes provided by wetlands.

Pollution is the silent enemy to standing waters. Smoke, contaminated runoff, spills, and other forms of pollution enter watersheds and harm aquatic organisms as well as those that feed upon them. Industrial waste, agricultural fertilizer runoff, and even human sewage can harm organisms and their habitats. Nitrogen, for example, used as a fertilizer for crops, can be carried by rainwater into ponds, creating algae blooms that use up huge amounts of dissolved oxygen, causing eutrophication. This often results in the elimination of pollution-sensitive species. Contaminants seeping into vernal pools can be even more destructive, as vernal pools typically have no outflow. Chemical changes in closed pools last for generations.

Recent legislation has taken aim at pollution. The goal of the Clean Water Act of 1972 was to protect all kinds of waters. And new stringent standards on sulfur emissions aim to reduce the harmful effects of burning high-sulfur coal, namely acid rain, which can be ruinous to the animals

Left: Water arums in shallow water bloom in August.

Right: Spatterdock is a common emergent species in Ohio ponds. Decomposed aquatic plants, along with sediment, are contributors to pond succession.

Fishing spiders (*Dolomedes* spp.) are adorned with hairs to detect vibrations. Like water striders, they can tread across the pond's surface. When fishing spiders dive underwater, air bubbles are trapped among their hairs, making them appear silver-colored.

in vernal pools. But recent court challenges have weakened the power of such laws. Ultimately, the careful stewardship of ponds and vernal pools will rest on private landowners, communities, and voters. Learning about aquatic ecology and biology and then participating in decision-making leads to the best form of environmental conservation.

Climate Change

The largest, most widespread threat to ponds and vernal pools may be climate change. We have come to understand that carbon emissions are not cost-free. As carbon dioxide builds in the atmosphere, climates around the world are changing. The earth is getting warmer. The exact consequences are not fully known, but scientists concur that climates around the world generally will get warmer, weather patterns will be altered, oceans will rise, and water in many areas will become more scarce. The effects in Ohio are not certain. Some ponds and pools may remain, but any increases in temperatures will likely affect many species that inhabit these vital waters.

What We Can Do

Stewardship of our natural resources, ponds and vernal pools included, depends upon individual involvement in local, regional, national, and

Opposite: In its last stage of succession, this pond has become a sedge meadow.

Left: This mayfly will live for only a few days. With only rudimentary mouthparts and no way to digest food, adult mayflies spend their energy on reproduction.

Right: Looking like it is decorated with glittering jewels, a dew-covered mosquito rests at the pond's edge.

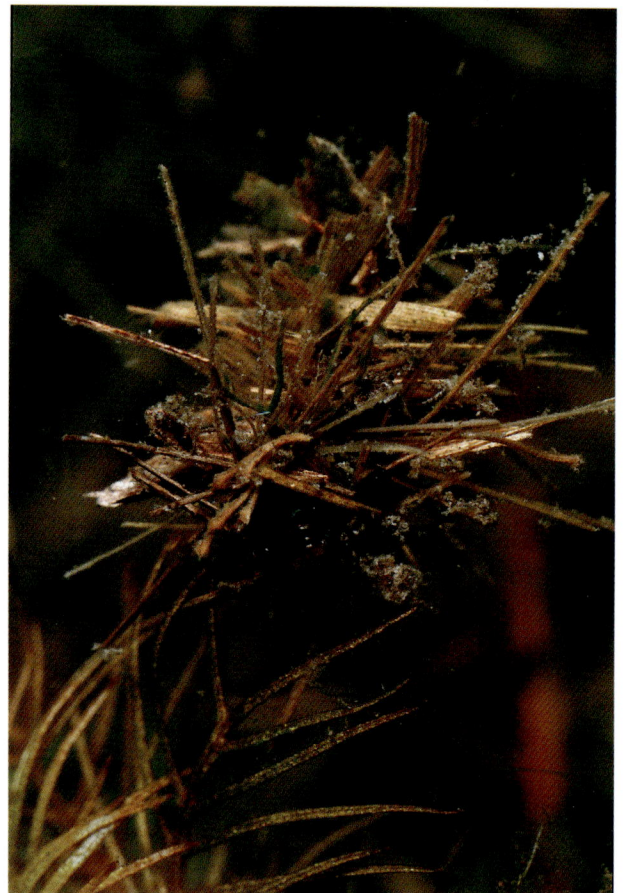

Left: Water scorpions are an aquatic version of the praying mantis. Water scorpions breathe through cylindrical tubes at the ends of their abdomens.

Right: Casemaker caddisflies become abundant as ponds, through succession, begin to fill in with sediment.

international politics. Management starts with individual decisions regarding the use of Earth's natural resources. The traditional three Rs—reduce, reuse, and recycle—are practical reminders of how individuals, neighborhoods, states, and nations can most efficiently utilize materials and energy. Understanding our world and then acting responsibly are the first steps toward lessening our impact. That is, part of the solution is the continued process of education, studying our natural worlds, sharing them with others, and then protecting them through all kinds of actions. Thus, the purpose of this book is simple: only by knowing and appreciating our natural world can we ever care enough to conserve it.

Bibliography

American Society of Mammalogists. http://www.mammalsociety.org/index.html.

Amos, William. *The Life of the Pond*. New York: McGraw-Hill, 1967.

Austin, Milton, Heidi Devine, Larry Goedd, Mike Greenlee, Tom Hall, Larry Johnson, and Paul Moser, eds. *Ohio Pond Management Handbook: A Guide to Managing Ponds for Fishing and Attracting Wildlife*. Columbus: ODNR–Division of Wildlife, 1996.

Batzer, Darold P., Russell B. Rader, and Scott A. Wissinger, eds. *Invertebrates in Freshwater Wetlands of North America: Ecology and Management*. New York: Wiley, 1999.

Bishop, Sherman C. *Handbook of Salamanders: The Salamanders of the United States, of Canada, and of Lower California*. Ithaca, NY: Comstock, 1967.

Borror, Donald Joyce, and Richard E. White. *A Field Guide to Insects: America North of Mexico*. Boston: Houghton Mifflin, 1998.

Bowers, Nora, Rick Bowers, and Kenn Kaufman. *Mammals of North America*. New York: Houghton Mifflin, 2004.

Boyd, Claude E. *Bottom Soils, Sediment, and Pond Aquaculture*. New York: Chapman & Hall, 1995.

Brinkley, Edward S. *National Wildlife Federation Field Guide to Birds of North America*. New York: Sterling, 2007.

Bronmark, Christer, and Lars-Anders Hansson. *The Biology of Lakes and Ponds*. 2nd ed. New York: Oxford University Press, 2005.

Bull, John L., and John Farrand, Jr. *The National Audubon Society Field Guide to North American Birds. Eastern Region*. New York: Knopf, 1996.

Caduto, Michael J. *Pond and Brook: A Guide to Nature in Freshwater Environments*. Hanover, NH: University Press of New England, 1990.

Calhoun, Aram J. K., and Phillip G. DeMaynadier, eds. *Science and Conservation of Vernal Pools in Northeastern North America: Ecology and Conservation of Seasonal Wetlands in Northeastern North America*. Boca Raton, FL: CRC Press, 2008.

Childs, Nancy, and Elizabeth Colburn. *Vernal Pool Lessons and Activities: A Curriculum Companion to* Certified: A Citizen's Step-By-Step Guide to Protecting Vernal Pools. Lincoln, MA: Massachusetts Audubon Society, 1993.

Chu, H. F. *How to Know the Immature Insects*. Dubuque, IA: William C. Brown, 1949.

Coe, James. *Eastern Birds: A Guide to Field Identification*, Revised and updated. New York: St. Martin's Press, 2001.

Colburn, Betsy, Tom Tyning, and Christopher Leahy, eds. *Pondwatchers Guide to Ponds and Vernal Pools of Eastern North America*. Illustrated by Barry W. Van Dusen. Lincoln, MA: Massachusetts Audubon Society, 1995.

Colburn, Elizabeth A., ed. *Certified: A Citizen's Step-by-Step Guide to Protecting Vernal Pools*. Edited by Lincoln, MA: Massachusetts Audubon Society, 1995.

———. *Vernal Pools: Natural History and Conservation*. Blacksburg, VA: McDonald and Woodward, 2004.

Conant, Roger. *A Field Guide to Reptiles and Amphibians of Eastern and Central North America*. 2nd ed. Illustrated by Isabelle Hunt Conant. Boston: Houghton Mifflin, 1975.

Cornell Lab of Ornithology and American Ornithologists' Union. *The Birds of North America Online*. http://bna.birds.cornell.edu/bna/.

———. *Reptiles of Ohio, 2nd ed*. Notre Dame, IN: University of Notre Dame Press, 1938.

Davis, Jeffrey G., and Scott A. Menze. *In Ohio's Backyard: Frogs and Toads*. Illustrated by Michael F. Wright and Jason Vineyard. Columbus: Ohio Biological Survey, 2002.

———. *Ohio Frog and Toad Atlas*. Columbus: Ohio Biological Survey, 2000.

Dexter, R. W., and C. H. Kuehnle. "Fairy Shrimp Populations of Northeastern Ohio in the Seasons of 1945 and 1946." *Ohio Journal of Science* 48 (1948): 15–26.

———. "Further Studies on the Fairy Shrimp Populations of Northeastern Ohio in the Seasons of 1945 and 1946." *Ohio Journal of Science* 51 (1948): 73–86.

Downer, Ann. *Spring Pool: A Guide to the Ecology of Temporary Ponds*. Boston: New England Aquarium, 1992.

Duellman, William Edward, and Linda Trueb. *Biology of Amphibians*. Illustrated by Linda Trueb. New York: McGraw-Hill, 1986.

Dunkle, Sidney. *Dragonflies through Binoculars: A Field Guide to Dragonflies of North America*. New York: Oxford, 2000.

Dunn, Jon L., and Jonathan Alderfer. *National Geographic Field Guide to the Birds of North America*. 5th ed. Washington, DC: National Geographic Society, 2006.

Eastman, John. *Birds of Lake, Pond, and Marsh: Water and Wetland Birds of Eastern North America*. Illustrated by Amelia Hansen. Mechanicsburg, PA: Stackpole Books, 1999.

———. *The Book of Swamp and Bog: Trees, Shrubs, and Wildflowers of the Eastern Freshwater Wetlands*. Illustrated by Amelia Hansen. Mechanicsburg, PA: Stackpole, 1995.

Eaton, Eric, and Kenn Kaufman. *Kaufman Field Guide to Insects of North America*. New York: Houghton Mifflin Harcourt, 2007.

Eddy, Samuel. *The Freshwater Fishes*. 3rd ed. Dubuque, IA: W. C. Brown, 1978.

Evans, Arthur V. *National Wildlife Federation Field Guide to Insects and Spiders and Related Species of North America*. New York: Sterling, 2007.

Farb, Peter. *The Insects: Life Nature Library*. New York: Time, 1962.

Frey, David G., ed. *Limnology in North America*. Madison: University of Wisconsin Press, 1966.

Glotzhober, Robert C., and Dave McShaffrey, eds. *The Dragonflies and Damselflies of Ohio*. Columbus: Ohio Biological Survey, 2002.

Gottschang, Jack L. *A Guide to the Mammals of Ohio*. Photography by Karl Maslowski et al. Illustrated by Elizabeth Dalve. Columbus, OH: Ohio State University Press, 1981.

Hamberger, John. *Birth of a Pond*. New York: Coward, McCann & Geoghegan, 1975.

Hausman, Leon. *Beginner's Guide to Fresh-Water Life*. New York: Putnam, 1950.

Insects on the Web. http://www.insects.org/.

Iowa State University Dept. of Entomology. BugGuide.net. http://www.bugguide .net.

Josephs, David. *Lakes, Ponds, and Temporary Pools*. New York: Franklin Watts, 2000.

Kays, Roland, and Don E. Wilson. *Mammals of North America*. Princeton, NJ: Princeton University Press, 2002.

Kenney, Leo P., ed. *Diving into Wicked Big Puddles*. Reading, MA: Vernal Pool Association, 1996.

Kenney, Leo P., and Matthew R. Burne. *A Field Guide to the Animals of Vernal Pools*. Westborough, MA: Massachusetts Division of Fisheries and Wildlife, 2001.

Lindeen, Carol. K. *Life in a Pond*. Mankato, MN: Capstone, 2004.

Marshall, Stephen Archer. *Insects: Their Natural History and Diversity, With a Photographic Guide to Insects of Eastern North America*. Richmond Hill, ON: Firefly, 2006.

McCafferty, W. Patrick. *Aquatic Entomology: The Fishermen's and Ecologists' Illustrated Guide to Insects and Their Relatives*. Illustrated by Arwin V. Provonsha. Boston: Jones and Bartlett, 1983.

McShaffrey, Dave, and Bob Glotzhober. *Dragonflies and Damselflies of Ohio: Field Guide*. Columbus: ODNR, Division of Wildlife, 2008.

Mead, Kurt. *Dragonflies of the North Woods*. Duluth, MN: Kollath-Stensaas Press, 2003.

Merritt, Richard W., and Kenneth W. Cummins, eds. *An Introduction to the Aquatic Insects of North America*. 3rd ed. Dubuque, IA: Kendall/Hunt, 1996.

Milne, Lorus Johnson, and Margery Milne. *The Audubon Society Field Guide to North American Insects and Spiders*. New York: Random House, 1980.

Morgan, Ann Haven. *Field Book of Ponds and Streams*. New York: G. P. Putnam's Sons, 1930.

Moyle, Peter B., and Joseph J. Cech. *Fishes: An Introduction to Ichthyology*. Upper Saddle River, NJ: Prentice Hall, 2003.

Ohio Department of Natural Resources. *Amphibians of Ohio Field Guide*. ODNR, Division of Wildlife Publication 348 (308). Columbus: ODNR, 2008.

———. *Mammals of Ohio*. Columbus: ODNR, Division of Wildlife, 2009.

———. *Ohio Fish Identification Guide*. Columbus: ODNR, Division of Wildlife, 2001.

———. *Reptiles of Ohio*. Columbus: ODNR, Division of Wildlife, 2008.

———. *Waterbirds of Ohio*. Columbus: ODNR, Division of Wildlife, 2006.

Ohio Frog and Toad Calling Survey and the Ohio Salamander Monitoring Program. OhioAmphibians.com. http://www.ohioamphibians.com/index.html.

Ohio Ornithological Society. http://www.ohiobirds.org/index.php.

Ohio Vernal Pool Partnership. http://www.ovpp.org/.

Page, Lawrence M., and Brooks M. Burr. *A Field Guide to Freshwater Fishes: North America North of Mexico*. Edited by Roger Tory Peterson. Illustrated by Eugene C. Beckham, Craig Wayne Ronto, and John P. Sherrod. New York: Houghton Mifflin, 1991.

Peckarsky, Barbara L., Pierre R. Fraissinet, Marjory A. Penton, and Don J. Conklin, Jr. *Freshwater Macroinvertebrates of North America*. Ithaca, NY: Cornell University Press, 1990.

Pennak, Robert W. *Fresh-Water Invertebrates of the United States, Protozoa to Mollusca*. New York: Wiley, 1989.

Peterjohn, Bruce G. *The Birds of Ohio*. Bloomington: Indiana University Press, 1989.

Peterson, Roger Tory. *A Field Guide to the Birds: A Completely New Guide to All the Birds of Eastern and Central North America*. 5th ed. New York: Houghton Mifflin, 2002.

———. *Peterson Field Guide to Birds of North America*. New York: Houghton Mifflin, 2008.

Petranka, James W. *Salamanders of the United States and Canada*. Washington: Smithsonian Institution Press, 1998.

Pfingsten, Ralph A., and Floyd L. Downs. *Salamanders of Ohio*. Columbus: Ohio State University, College of Biological Sciences, 1989.

Pfingsten, Ralph A., and Timothy O. Matson. *Ohio Salamander Atlas*. Columbus: Ohio State University, College of Biological Sciences, 2003.

Redington, Charles B. *Plants in Wetlands*. Dubuque, IA: Kendall/Hunt, 1994.

Reid, Fiona. *Peterson Field Guide to Mammals of North America*. Boston: Houghton Mifflin, 2006.

Reid, George K. *Pond Life*. Revised and updated. Illustrated by Sally D. Kaicher and Tom Dolan. New York: St. Martin's Press, 2001.

The Reptile Information Network. http://www.reptileinfo.com/Home.aspx.

Rosche, Larry, Judy Semroc, and Linda Gilbert. *Dragonflies and Damselflies of Northeast Ohio*. 2nd ed. Illustrated by Jennifer Brumfield. Cleveland: Cleveland Museum of Natural History, 2008.

Russell, Franklin. *Watchers at the Pond*. Illustrated by Robert W. Arnold. New York: Time, 1966.

Schmidt, Bob. *Advanced Sport Fishing and Aquatic Resources Handbook: Ohio*. Illustrated by John Rice. Dubuque, IA: Kendall/Hunt, 1997.

———. *Sport Fishing and Aquatic Resources Handbook: Ohio*. Illustrated by John Rice. Dubuque, IA: Kendall/Hunt, 1991.

Sibley, David Allen. *The Sibley Field Guide to Birds of Eastern North America*. New York: Chanticleer, 2003.

———. *The Sibley Guide to Birds*. New York: Knopf, 2000.

Smith, Douglas Grant. *Pennak's Freshwater Invertebrates of the United States: Porifera to Crustacea*. 4th ed. New York: Wiley, 2001.

Smith, Hobart Muir. *Amphibians of North America: A Guide to Field Identification*. Illustrated by Sy Barlowe. New York: Golden Press, 1978.

Smith, Hobart M., and Edmund D. Brodie, Jr. *Reptiles of North America*. New York: Golden, 1982.

Stewart, Ray. "Fairy Shrimp: The Ephemeral Enigma." Friends of Wetlands Newsletter (February 2006), http://www.oupp.org/CreatureFeature.html.

Tekiela, Stan. *Birds of Ohio*. Cambridge, MN: Adventure, 1999.

Thoma, Roger F., and Raymond F. Jezerinac. *Ohio Crayfish and Shrimp Atlas*. Columbus: Ohio Biological Survey, 2000.

Thompson, Bill, III. *The Young Birder's Guide to Birds of Eastern North America*. Illustrated by Julie Zickefoose. New York: Houghton Mifflin, 2008.

Thorp, James H., and Alan P. Covich, eds. *Ecology and Classification of North American Freshwater Invertebrates*. San Diego: Academic Press, 2001.

Tiner, Ralph, Jr. *Field Guide to Nontidal Wetland Identification*. Annapolis, MD: Maryland Department of Natural Resources and U.S. Fish and Wildlife Service, 1988.

Trautman, Milton Bernhard. *The Fishes of Ohio*. Columbus: Ohio State University Press, 1957.

Tyning, Thomas F. *A Guide to Amphibians and Reptiles*. New York: Little, Brown, 1990.

The Vernal Pool Association. http://www.vernalpool.org/vernal_1.htm.

Voshell, J. Reese. *A Guide to Common Freshwater Invertebrates of North America*. Illustrated by Amy Bartlett Wright. Blacksburg, VA: McDonald & Woodward, 2002.

Waldbauer, Gilbert. *A Walk around the Pond: Insects in and over the Water*. New Haven: Harvard University Press, 2006.

Ward, J. V. *Aquatic Insect Ecology. Vol. 1, Biology and Habitat*. New York: Wiley, 1992.

Westfall, Minter Jackson. *Damselflies of North America*. Gainesville, FL: Scientific Publishers, 1996.

Wetzel, Robert G. *Limnology: Lake and River Ecosystems*. 3rd ed. New York: Academic Press, 2001.

Whitaker, John, Jr. *National Audubon Society Field Guide to North American Mammals*. New York: Knopf, 1997.

Wright, Anna Allen, and Albert Hazen Wright. *Handbook of Frogs and Toads of the United States and Canada*. Ithaca, NY: Comstock, 1949.

Wynn, Douglas E., and Scott M. Moody. *Ohio Turtle, Lizard, and Snake Atlas*. Columbus: Ohio Biological Survey, 2006.

Zim, Herbert S., and Clarence Cottam. *Insects*. Revised and updated. Illustrated by James Gordon Irving. New York: St. Martin's Press, 2001.

Zimmer, Carl. "Walking on Water." *Natural History* 109, no. 3 (April 2000): 30.

Recommended Children's Books

Becker, Julie. *Animals of the Ponds and Streams*. Illustrated by Maarja Roth-Evenson. St. Paul, MN: EMC, 1977.

Coatsworth, Elizabeth Jane. *Under the Green Willow*. Illustrated by Janina Domanska. New York: Macmillan, 1971.

Cole, Henry. *I Took a Walk*. New York: Greenwillow, 1998.

Conant, Roger, Robert C. Stebbins, and Joseph T. Collins. *Peterson First Guide to Reptiles and Amphibians*. Boston: Houghton Mifflin, 1992.

Ellis, Brian. *The Web at Dragonfly Pond*. Illustrated by Michael S. Maydak. Nevada City, CA: Dawn, 2006.

Falwell, Cathryn. *Turtle Splash! Countdown at the Pond*. New York: Greenwillow, 2008.

Fleisher, Paul. *Lake and Pond Food Webs*. Minneapolis: Lerner, 2008.

———. *Pond*. New York: Benchmark, 1999.

Fleming, Denise. *In the Small, Small Pond*. New York: Henry Holt, 1993.

Fraser, Mary Ann. *Where Are the Night Animals?* New York: Scholastic, 1999.

Fredericks, Anthony D. *Near One Cattail: Turtles, Logs, and Leaping Frogs*. Illustrated by Jennifer DiRubbio. Nevada City, CA: Dawn, 2005.

Ganeri, Anita. *From Tadpole to Frog*. Chicago: Heinemann, 2006.

George, Lindsay Barrett. *Around the Pond: Who's Been Here?* New York: Greenwillow, 1996.

Hibbert, Adam. *A Freshwater Pond*. New York: Crabtree, 1999.

Himmelman, John. *A Salamander's Life*. New York: Children's Press, 1998.

Hunter, Anne. *What's in the Pond?* New York: Houghton Mifflin, 1999.

Johnson, Jinny. *Simon & Schuster Children's Guide to Insects and Spiders*. New York: Simon & Schuster, 1996.

Kavanagh, James. *Pond Life Nature Activity Book: Educational Games and Activities for Kids of All Ages*. Illustrated by Raymond Leung. Phoenix: Waterford, 2002.

Kirkpatrick, Rena K. *Look at Pond Life*. Illustrated by Annabel Milne and Peter Stebbing. Milwaukee: Raintree Children's Books, 1978.

Koch, Maryjo. *Pond Lake River Sea*. San Francisco: Collins, 1994.

Kottke, Jan. *From Tadpole to Frog*. New York: Scholastic, 2001.

Lindeen, Carol. *Life in a Pond*. Mankato, MN: Capstone, 2004.

Maruska, Edward J. *Salamanders*. Chanhassen, MN: Child's World, 2006.

Mazer, Anne. *The Salamander Room*. Illustrated by Steve Johnson. New York: Knopf, 1991.

McKeever, Susan. *Freshwater Life of North America*. Illustrated by Colin Newman. San Diego: Thunder Bay, 1995.

McNab, Chris. *Frogs, Toads and Salamanders*. Milwaukee: Gareth Stevens, 2006.

Merrifield, Susan. *Fishing ABC's Coloring Book*. Illustrated by Mat Merchlewitz. Atlanta, GA: U.S. Fish and Wildlife Service, 1994. http://www.dnr.state.oh.us/wildlife/PDF/fishcolorbook.pdf.

Michels, Tilde. *At the Frog Pond*. New York: Lippincott, 1989.

Miller, Sara Swan. *Salamanders: Secret, Silent Lives*. Illustrations by Jose Gonzales and Steve Savage. New York: Franklin Watts, 1999.

Morgan, Sally. *From Tadpole to Frog*. North Mankato, MN: Thameside, 2002.

Morrison, Gordon. *Pond*. New York: Houghton Mifflin, 2002.

Parker, Steve. *Pond and River*. New York: Dorling Kindersley, 2005.

Pfeffer, Wendy. *From Tadpole to Frog*. Illustrated by Holly Keller. New York: HarperCollins, 1994.

Pratt-Serafini, Kristin Joy. *Salamander Rain: A Lake and Pond Journal*. Nevada City, CA: Dawn, 2000.

Rockwell, Anne. *Ducklings and Pollywogs*. Illustrated by Lizzy Rockwell. New York: Macmillan, 1994.

Schofield, Jennifer. *Animal Babies in Ponds and Rivers*. Boston: Kingfisher, 2004.

Sidman, Joyce. *Song of the Water Boatman and Other Pond Poems*. Illustrated by Beckie Prange. Houghton Mifflin, 2005.

Sill, Cathryn P. *About Mammals: A Guide for Children*. Illustrated by John Sill. Atlanta: Peachtree, 1999.

Silver, Donald M. *Busy Beaver Pond: A Pop-Up Book*. Illustrated by Patricia J. Wynne. New York: W. H. Freeman, 1995.

———. *Pond*. Illustrated by Patricia J. Wynne. New York: Learning Triangle, 1997.

Stewart, David. *From Tadpole to Frog*. Illustrated by Carolyn Scrace. New York: Franklin Watts, 1998.

Tafuri, Nancy. *Goodnight, My Duckling*. New York: Scholastic, 2005.

———. *Have You Seen My Duckling?* New York: Greenwillow, 1996.

Taylor, Barbara. *Pond Life*. Photography by Frank Greenaway. London: Coven Garden Books, 1998.

Waddell, Martin. *Pig in the Pond*. Cambridge, MA: Candlewick, 1996.

Wechsler, Doug. *Frog Heaven: Ecology of a Vernal Pool*. Honesdale, PA: Boyds Mills, 2006.

Wong, Herbert H., and Matthew F. Vessel. *Pond Life: Watching Animals Find Food*. Illustrated by Tony Chen. Reading, MA: Addison-Wesley, 1970.

Zemlicka, Shannon. *From Tadpole to Frog*. Minneapolis: Lerner, 2003.

Zoehfeld, Kathleen Weidner. *From Tadpole to Frog*. Photographs by Dwight Kuhn. New York: Scholastic, 2001.

Scientific Supply Companies

Carolina Biological Supply—Wide range of science supplies; good insect nets.
   http://www.carolina.com
Frey Scientific—Wide range of science supplies.
   http://www.freyscientific.com
LaMotte—Water-testing kits.
   http://www.lamotte.com
Memphis Net and Twine—Heavy-duty dip nets.
   http://www.memphisnet.net
Ward's Natural Science—Wide range of science supplies.
   http://www.wardsci.com

Glossary

Acoustico-Lateral Hearing System—secondary hearing system in fish, most amphibian larvae, and some amphibian adults, in which the *neuromasts*—tiny sound sensors in a line along each side of the body, the *lateral line*—perceive extremely low-frequency sounds.

Aerobic—containing oxygen.

Algae—single or multicelled *autotrophic* plants, usually found in water and lacking true roots, stems, or leaves.

Allochthonous—when an ecosystem's energy is produced *photosynthetically* elsewhere and imported (from the Greek *allos*, "other" and *khthon*, "earth").

Amplexus—the mating position of frogs and toads, where the female lays eggs while the male, on her back and "hugging" her with his forelimbs, fertilizes the eggs (from the Latin *amplexus*, "to embrace").

Autochthonous—generally, native; in limnology, when an aquatic ecosystem's energy is produced *photosynthetically* from within itself (from the Greek *autos*, "self" and *khthon*, "earth").

Autotroph—an organism, such as a plant, that is capable of utilizing carbon dioxide (found in water and air) and environmental energy (usually sunlight) to produce the organic molecules necessary for living.

Bioluminescence—the production of light by an organism through a chemical reaction.

Bivalves—another name for clams and mussels (class *Pelecypoda*), derived from the use of two siphons in breathing and feeding.

Borrow Pit—a hole left in earth where material was taken or "borrowed" for use elsewhere, such as along roadways to build up overpasses.

Buffering—the neutralizing of acids by bases, such as calcareous bedrock neutralizing acid rain.

Bufotenins—toxic chemicals released by the *parotid glands* of true toads.

Buteo—medium-sized raptors with broad wings, large bodies, and wide tails that aid in soaring.

Carapace—in turtles, the upper portion of the shell.

Carnivore—a *heterotrophic*, consumer organism that eats other consumers, usually an animal that eats other animals.

Cephalothorax—a single body segment comprised of the head and thorax.

Channel Ponds—ponds formed in old river beds, such as abandoned oxbows.

Chelicerae—fanglike appendages on arachnids used for grasping and piercing.

Chitin—a protein material (polysaccharide) that forms the exoskeleton of animals such as crayfish and insects.

Chloroplasts—organelles, found within producers, that contain *chlorophyll*.

Chlorophyll—green pigment, found in plants, *algae*, and cyanobacteria, which absorbs sunlight and allows for the conversion of carbon dioxide, water, and sunlight into glucose and oxygen, a process known as *photosynthesis*.

Cloaca—a chamber in lower vertebrates that collects and releases materials from the intestine, urinary bladder, and sex organs (e.g., urine, feces, sperm, and eggs).

Colubrid—snakes in the family *Colubridae*, which comprises approximately two-thirds of all snake species.

Consumer—a *heterotrophic* organism (typically an animal) that eats other organisms, including producers, decomposers, and other consumers.

Crest—showy feathers growing on the heads of birds.

Decapods—an order of crustaceans comprised of crayfish and shrimp.

Decomposer—a *heterotrophic* organism, such as a fungus or bacterium, that derives energy from dead or waste organic material, causing the breakdown of organic substances into inorganic compounds.

Detritus—nonliving organic matter, either in dissolved form (< 1/50,000 of an inch in diameter) or particulate form (> 1/50,000 of an inch in diameter).

Detrivores—a heterotrophic organism that ingests detritus.

Dextral—meaning "on the right," the term is used to describe the side from which an elongate gastropod's shell opening faces when the narrow end is up and the *operculum* is facing the viewer.

Diapause—a period of rest, especially useful for organisms to survive inhospitable conditions, such as droughts.

Dissolved Organic Matter (DOM)—organic matter of less than 1/50,000 of an inch in size.

Dorsal—referring to the upper surface; as opposed to *ventral*.

Dorsal Fin—the fin located on a fish's back or top.

Dorsolateral Fold—a fold of skin on frogs that separates the back from the side.

Drainage Basin—the area from which a body of water derives its inflow.

Ecosystem—the communal interaction of organisms with other organisms and with their shared physical environment through the flow of energy and the cycling of materials.

Eimer's Organs—sensory organs found on the noses of moles, such as the star-nosed mole.

Elytron—(plural *elytra*) hardened front wings that fold over the rear wing and abdomen of beetles (order *Coleoptera*), true bugs (order *Hemiptera*), and other insects, providing protection and, for some aquatic beetles, trapping air bubbles for underwater breathing.

Emergent Vegetation—plants with submerged roots whose stalks rise above the surface of the water.

Endoskeleton—a supporting structure on the inside of an animal, as opposed to *exoskeleton*.

Eutrophication—the process of overenriching bodies of water with nutrients, thus resulting in excessive growth of *algae* and *macrophytes* and in oxygen depletion; the reduction in dissolved oxygen leads to the death of aquatic plants and animals.

Evaporation—the process of water changing from a liquid state to a gaseous state via the input of energy.

Exoskeleton—a supporting structure on the outside of an animal, as opposed to *endoskeleton*.

Extirpation—the complete removal of a species from an area, from the Latin for "to uproot."

Exuviae—*exoskeletons* left behind by molting arthropods; such "skins" are often left on pond vegetation by odonates *metamorphosing* from larvae to adults.

Facultative Species of Vernal Pools—species that can be found both in vernal pools and in other wetland, lacustrine, or riverine environments.

Fossorial—adapted to burrowing.

Freshwater—nonsalinated water, as opposed to *seawater*.

Fry—newly hatched fish able to actively feed (i.e., no longer drawing from their yolk sacs).

Food Chain—a simplified model of consumption within an *ecosystem*, beginning with *producers* (typically plants) and continuing with *consumers* (typically animals), each of whom feed upon the previous organism in the chain; the chain continues, with dead material and waste products being recycled (typically by *decomposers*) back to the primary chemical compounds necessary for producers (again plants) to start the sequence over again.

Groundwater—water within the earth, either in streams or aquifers.

Halteres—also known as balancers or poisers, these modified wings, which are short stalks with knobs on the end and are located behind the front wings of diptera, or flies, aid in stability during flight.

Hamules—second genitalia of *odonates*, located on the underside and within the second abdomen section, into which males place their sperm for copulation.

Hellgrammites—larvae of dobsonflies and fishflies.

Herbivore—a *heterotrophic* organism or *consumer* that eats *producers*, usually an animal that eats plants.

Hermaphroditic—possessing both male and female reproductive organs.

Heterotroph—an organism dependent upon autotrophs to supply the carbon and energy necessary for living (e.g., an animal that eats plants).

Hydrology—the scientific study of water and its distribution and movement, including sources and patterns of changing levels.

Hydroperiod—the length of time that water is present in a wetland.

Icterids—members of the family *Icteridae*, which are small and often colorful passerine birds and include blackbirds, grackles, orioles, and cowbirds.

Invertebrate—an animal lacking a vertebral column.

Insectivore—a *carnivorous* organism whose diet consists of insects.

Labium—one of the mouthpart structures in insects, namely, the lower lip.

Lateral Line—part of a fish's *acoustico-lateral hearing system*, the lateral line runs down each side of the fish's body and is made up of tiny *neuromasts,* specialized hair cells that perceive vibrations, not only used in locating food but also hypothesized to aid in maneuvering within schools and sensing changes in the weather.

Lacustrine—of or pertaining to lakes.

Lentic—referring to standing, not moving, water; e.g., a pond is lentic, as opposed to river or stream, which is *lotic.* Lentic is derived from the Latin word *lenis,* meaning "gentle."

Lotic—referring to moving, not standing, water; for example, a stream is lotic, as opposed to a pond, which is *lentic.* Lotic is derived from the Latin *lavare,* meaning "to wash."

Macroinvertebrates—invertebrates able to be seen without a microscope.

Macrophytes—large plants (i.e., nonmicroscopic plants).

Malacostraca—large, bottom-dwelling crustaceans, such as crayfish, shrimp, aquatic sow bugs, and side swimmers.

Marginal Plates—the bony plates surrounding the lower edge of a turtle's *plastron.*

Marsupium—external pouch on the *venter* that is used for holding offspring.

Medusa—the bell-shaped, freely swimming, and sexually mature stage of a *cnidarian.*

Metamorphosis—the process of changing body form and structures from one stage of life to another, as seen in insects and amphibians.

Microinvertebrates—invertebrates requiring a microscope to be seen.

Molt—the shedding of an invertebrate's skin, allowing for growth.

Nares—nostrils of fish, typically four, behind which lie highly sensitive odor detectors.

Neuromasts—tiny sound sensors, able to perceive low-frequency sounds, found in the lateral line of fish.

Nocturnal—active at night.

Nymph—the second stage of development of insects that undergo incomplete *metamorphosis,* such as dragonflies and damselflies.

Obligate Species of Vernal Pools—species whose survival depends upon periodic flooding and drying and that must be present within a temporary pool in order for the wetland to classify as a *vernal pool.*

Odonate—any dragonfly or damselfly, of the order *Odonata.*

Omnivore—a organism that is both *herbivorous* and *carnivorous.*

Opercle—the "ear flap" or bony covering protecting the gills on a fish.

Operculum—the flat, hard plate protecting a gastropod's soft tissue at the shell's opening.

Organism—a living being from any of the seven kingdoms (*Archaebacteria, Eubacteria, Protista, Chromista, Plantae, Animalia,* and *Eumycota*).

Organic Matter—a carbon-based compound that is created by organisms; most carbon-based compounds are organic compounds, with the exception of carbon monoxide, carbon dioxide, and a number of carbon-containing inorganic salts.

Outwash—material, usually sand or gravel, carried by water from a melting glacier.

Parotid glands—prominent glands right behind the eyes of true toads, which secrete toxic chemicals called *bufotenins.*

Particulate Organic Matter (POM)—organic matter greater than 1/50,000 of an inch in size.

Photolysis—the breaking down of chemical compounds into their constituent parts due to sunlight, typically from interactions with photons of ultraviolet light.

Photonegative—moving away from or avoiding light.

Photosynthesis—the autotrophic process of converting carbon dioxide, water, and energy from sunlight, in the presence of *chlorophyll,* into glucose and oxygen ($6CO_2 + 6H_2O$ Sunlight$\rightarrow C_6H_{12}O_6 + 6O_2$).

Physical Gill—bubble of air carried by some true bugs and beetles that acts as an indefinite store of oxygen. As respiration occurs, dissolved oxygen in the water diffuses into the bubble, maintaining a continuous level of twelve percent. Similarly but in reverse, carbon dioxide released by the insects diffuses into the surrounding water.

Phytoplankton—minute, mostly microscopic, freely drifting plants and photosynthetic bacteria (some with very limited powers of locomotion) that are suspended in water, often found only in *lentic* waters.

Planarian—a macroscopic invertebrate of the Phylum *Platyhelminthes,* class *Turbellaria,* order *Tricladida,* ranging in size from one-quarter inch to over one inch and typically earth-toned, flat, and soft with two "cross-eyed" eye spots on top.

Plankton—minute, mostly microscopic, freely drifting organisms with limited powers of locomotion that are suspended in water; either plant (*phytoplankton*) or animal (*zooplankton*).

Plastron—for aquatic insects, a thin film of air used as a *physical gill;* for turtles, the lower portion of the shell.

Pleopods—appendages, less technically called *swimmerets,* on decapods such as crayfish, used primarily for swimming but also employed by females for carrying their eggs.

Pond—a permanent body of standing water of relatively uniform temperature and both small enough that waves do not form and shallow enough to *potentially* allow light penetration across the entire basin clear to the bottom, thus allowing for the possibility of plant growth at the deepest point.

Producer—an *autotrophic* organism that produces complex organic compounds from simple inorganic compounds, typically through *photosynthesis*; that is, an organism able to produce its own food from sunlight.

Proleg—a fleshy, abdominal leg found on some insect larvae, such as the "feet" on caterpillars.

Radula—tonguelike projection, which is covered with sharp barbs and dragged back into a gastropod's mouth, scraping *algae* and collecting *detritus* and small invertebrates for feeding.

Raptor—bird of prey with sharp talons and hooked beaks.

Riverine—of or pertaining to rivers and streams.

Scutes—individual plates comprising the shell of a turtle.

Shredders—primary *consumers* who graze by ripping apart plant material into tiny pieces.

Sinistral—meaning "on the left," the term is used to describe the side from which an elongate gastropod's shell opening faces when the narrow end is up and the *operculum* is facing the viewer.

Siphon—in a mollusk, tubelike structure that is lined with undulating cilia that move water in and out of the bivalve's shell for breathing and feeding.

Speculum—a patch on the wings of birds and other animals, often metallic in color.

Spermataphore—a packet of male reproductive cells, or sperm, which are dropped by males and picked up by females for fertilization before egg laying.

Stigma—colored spot near the tip of clear-winged insects, such as dragonflies, damselflies, and bees.

Stridulation—the rubbing together of body parts to create sound.

Subimago—stage in mayfly development between larva and adult, when the insect is winged but not sexually mature.

Sublimation—the process of water changing from a solid state directly to a gaseous one via the input of energy.

Succession—the predictable changes effected upon ecological communities; in the case of ponds and pools, the natural filling-in process from sedimentation and deposition of organic matter.

Surface Inflow—water reaching a pond, pool, or other body of water through surface runoff from within the watershed.

Swimmerets—paddlelike appendages, more technically called *pleopods*, found on the abdomen of crayfish and other crustaceans used for locomotion and reproduction.

Tandem—the mating and egg-laying position of insects, in which the male grabs and holds the female.

Temporary Pool—a nonpermanent body of standing water of relatively uniform temperature, periodically filling and drying up.

Till—unstratified material deposited by a glacier.

Transpiration—the movement of water absorbed by a plant's roots and released into the air as vapor.

Triclad—see planarian.

Turbidity—the clouding of water by particulate or suspended matter, either organic or inorganic, causing coloration of the water and decreasing the transmission of light.

Vector—a physical carrier of a disease, such as mosquitoes that spread malaria.

Venter—lower surface (Latin for "belly"); ventral is the opposite of *dorsal*.

Vernal Pool—a specialized temporary pool that is utilized for reproduction by certain *obligate species*, such as mole salamanders (*Ambystoma* spp.) and wood frogs (*Rana sylvatica*).

Water Quality—the physical, chemical, and biological properties of water, which affect its use by humans and other organisms.

Wetland—an area that is inundated with water at least part of the year.

Wheel Position—mating position of *odonates*, where the male uses the claspers at the end of his abdomen to hold the female between her head and thorax while she brings her abdomen forward and up to pick up sperm from the male.

Zooplankton—minute, mostly microscopic, freely drifting animals (some with limited powers of locomotion), which are suspended in water.

Index

mussels, 42–43
planarians, 38
predators of, *58*, 64, 67, 69, 85, 87, 99, 105–6, 111
segmented worms, 39
shrimp, 46–47, 77
side swimmers, 49
six-spotted fishing spiders, 44–45
spiders, 43–45, 85, *123*
sponges, 51
worms, 38–39, 67, 85, 87, 99, 105

kettle holes, 4

Lake County, vernal pools in, 28
Lake Erie, 3–4
insects along, 62, 71
ponds formed by sand dunes along, 5–7
Lake Kelso, glacial formation of, 5
lakes, *vs.* ponds, 2–3
leaves, as source of allochthonous energy, 11
lentic water, definition of, 1
light, 3
detritus decomposed by ultraviolet, 14
factors limiting penetration of, 9, 15, 30
lotic water, 1, 37

macrophytes, 29, 34
mammals
as pond residents *vs.* visitors, 103
bats, 107
beavers, 7, *102*, 103–4, *105–6*
big brown bats, 107
diet of, 43, 85, 106
little brown bats, 107
minks, 106–7
muskrats, 43, 104, *104*
predators of, 40, 69, 85, 101, 106, 113, 114
raccoons, 43, 107
river otters, 105–6
rodents, 85
silver-haired bats, 107
star-nosed mole, 104–5
microorganisms
in vernal pools, 25, 28, 34
macroinvertebrates *vs.*, 38
microscope, to study pondlife, 119

migration
by dragonflies, 56–57
in reproductive strategies of obligate species of vernal pools, 32–33

nutrient cycles, in vernal pools, 34–35

Oak Opening, dune ponds in, 7
obligate species, of vernal pools, 24–25, 32–33
odonates. *See* damselflies; dragonflies
Ohio's Pond Management Handbook, 74–75
open ponds, *vs.* closed, 9, 122
organic compounds, 14
introduced as detritus, 10–12, 34
solar energy converted into, 10–11
outflow, 8
from ponds, 22
from vernal pools, 22–24, 122
oxygen
aquatic earthworms increasing dissolved, 39–40
mayfly larvae obtaining, 63
salamanders obtaining, 94, 97
supplying young with, 46, 47, 49
turtles obtaining, 99
water bugs obtaining, 60–61, 67–68
oxygen depletion, 70
from algae blooms, 9, 122
in vernal pools, 30

parasites, 39, 44
permanence
in definition of ponds, 2, 20
in definition of vernal pools, 20–23
photolysis, 14, 30. *See also* light
photosynthesis, 12
autochthonous energy production through, 10–11
in vernal pools, 30, 34
snails improving by eating algae, 41, *42*
phytoplankton, in food webs, 74
plankton, 62, 74

plants, 35. *See also* algae
bladderwort, *48*
buttonbush, 32
cattails, 32
consumers of, *45*, 49, 62, 67, 80–81, 99, 104, 109, 112–13
duckweed, *2*, *13*
effects on vernal pools' hydrology, 29–30
filling in ponds and vernal pools, 121
in defining lakes *vs.* ponds, 3
in energy cycles, 10–11, 62
in pond management, 74, 80–81
in vernal pools, 22, 26, 30–32
invasive, *118*
light penetration and, 3–4, 15
muskrats decimating, 104, *104*
pond lilies, 32
purple pickerelweed, *12*
spatterdock, *122*
surface, 15
swamp loosestrife, *118*
water arums, *122*
pollution, 123
chemical contamination as, 9, 30–31
effects of, 15, 30, 122
pond formation
by beavers, 7
by humans, 7, *10*, 73
glaciers in, 4–5, 26
importance of shape, 74
sand dunes in, 5–7
waterways in, 5
pond management, 73–75, 121
pond succession, 121, *122*, *124*, *126*
ponds, 53, 73, 88, 106
autochthonous *vs.* allochthonous production in, 32–34
cycles of life in, 119–20
definition of, 1–4, 20–22
purposes of, 7
threats to, 121–23
precipitation
acidity of, 30
as water source for ponds, 7–8
as water source for vernal pools, 22, 26, 28
producers, 14–15

rabies, among bats, 107
reptiles, 83. *See also* snakes; turtles
in vernal pools, 25, 33
predators of, 69, 114
Richland County Park District, formation of vernal pools by, 27
rivers, 5, 26
runoff, surface, 7, 22, 26, 34

salamanders
blue spotted, 96
breeding in vernal pools, 17–18, 28, 121
diet of, 34, 40, *97*
eastern tiger, 96, *97*
eggs of, 22, 64
four-toed, *24*, 97–98
hybridization among, 96
in vernal pools, 18–19, 22, 32–33, 94
indicator species of, 94
Jefferson, 17, 95
larvae of, *21*, 24–25, 27, 29, 34, 54, 62, 95, 97, 121
lungless, 96–97
marbled, 29, 34, 95, 96
mating of, 94
mole, 17–18, 22, 32–33, 93–94
newts, *1*, 22, 40, 82, 98, *98*
predators of, 22, 34, 64, 67, 101
red-spotted newts, *1*, 22, 82, 98, *98*
small-mouthed, *21*, 95–96
spotted, 17, *19*, 22, 25, 27, 94
tiger, 17
toads and frogs *vs.*, 93
sand dunes, in pond formation, 5–7
scavengers, water beetles as, 65, 67
Scioto River, vernal pools along, 27
seasons, 15, 120
sediment, filling in ponds and vernal pools, 121
seepage
from ponds, 8
from vernal pools, 22, 28–29
groundwater as source for pools and ponds, 7–8, 22
shredders, eating fallen leaves in vernal pools, 34